IRON & STEEL

Alabama

THE FORGE OF HISTORY

A SERIES OF ILLUSTRATED BOOKS

IRON & STEEL

A Guide to Birmingham Area Industrial Heritage Sites

James R. Bennett
and Karen R. Utz

The University of Alabama Press • Tuscaloosa

Manufactured in China
Typeface: AGaramond

∞
The paper on which this book is printed meets the minimum requirements of American National Standard for Information Sciences-Permanence of Paper for Printed Library Materials,
ANSI Z39.48-1984.

Library of Congress Cataloging-in-Publication Data

Bennett, James R. (James Ronald), 1940–
Iron & steel : a guide to Birmingham area industrial heritage sites / Jim Bennett and Karen Utz.
p. cm. — (Alabama : the forge of history)
Includes bibliographical references.
ISBN 978-0-8173-5611-8 (paper : alk. paper) — ISBN 978-0-8173-8398-5 (electronic) 1. Iron-works—Alabama—Birmingham Region—Guidebooks. 2. Coke-ovens—Alabama—Birmingham Region—Guidebooks. 3. Blast furnaces—Alabama—Birmingham Region—Guidebooks. 4. Parks—Alabama—Birmingham Region—Guidebooks. 5. Historic sites—Alabama—Birmingham Region—Guidebooks. 6. Birmingham Region (Ala.)—Guidebooks. 7. Industries—Alabama—Birmingham Region—History. 8. Industrial archaeology—Alabama—Birmingham Region. 9. Birmingham Region (Ala.)—History, Local. I. Utz, Karen R. II. Title. III. Title: Iron and steel.
TN704.U52A23 2010
669'.109761781—dc22

2010003232

The University of Alabama Press wishes to gratefully acknowledge the Office of the Provost of The University of Alabama for their generous support of this book.

Frontispiece: Tannehill Ironworks, 2010 snowfall, Hillman's Forge site is the center structure in the foreground. (Stacey Green.)

Contents

Illustrations

Figures

Maps

Acknowledgments

A book is never a singular undertaking. Writers and researchers alike lean on the work of many others, both in the present and the past.

To all those who contributed to this guidebook of Birmingham District industrial sites from whatever decade we offer our sincerest appreciation.

This effort drew on the works of many who have preceded us: Dr. W. David Lewis, professor emeritus at Auburn University and author of *Sloss Furnaces and the Rise of the Birmingham District* (1994); Joseph H. Woodward II, formerly of the Woodward Iron Company and author of *Alabama Blast Furnaces* (1940); and Ethel Armes, the noted Birmingham historian who penned *The Story of Coal and Iron in Alabama* (1910).

We also acknowledge present-day researchers including Marjorie White, director of the Birmingham Historical Society and author of *The Birmingham District: An Industrial History and Guide,* and Yvonne Crumpler, former director of the Southern History Department, Birmingham Public Library.

We are further grateful to the following institutions and organizations: Tannehill Ironworks Historical State Park, the Iron & Steel Museum of Alabama, Sloss Furnaces National Historic Site, the Alabama Historic Ironworks Commission, Birmingham-Jefferson Historical Society, Ruffner Mountain

Nature Center, the Red Mountain Greenway and Recreational Area Commission, and The University of Alabama Press.

A special thank you to Kate Lorenz, who kept our feet to the fire.

Introduction

The Birmingham Iron and Steel District has been one of the most historically important iron-producing centers in the United States.

During the Civil War, a half dozen early iron furnaces located in the metropolitan area not only invited federal invasion but heralded major postwar investment in the iron and steel industry. Closely associated deposits of iron ore, limestone, and coal enabled foundry-grade iron to be made here more cheaply

The Alice Furnaces, circa 1889. Its daily output for its first year of operation in 1880 was 53 tons of foundry iron, a record at that time. A second furnace shown here was added in 1883. (John Horgan Jr. Collection, Hoole Special Collections Library, University of Alabama.)

Birmingham's Ensley Works, 1909. (Haines Photo Co., Library of Congress.)

than anywhere else in the nation. By 1874, the cost of making iron ore was quoted at $20 per ton.

Ruins of nineteenth-century furnaces at Tannehill, Brierfield, Shelby, and Irondale speak to the laborious ways of making iron in the 1800s. All these sites are currently incorporated into public parks. Downtown, the Sloss City Furnaces are now a part of Sloss Furnaces National Historical Landmark.

The Birmingham District also features numerous iron ore and coal-mining sites, remains of coke ovens, and other historic places tied to the iron trade. From Tannehill, which began making iron in 1830, to the Sloss City Furnaces, which helped make Birmingham the "Pittsburgh of the South," history buffs and tourists alike can walk the very pathways of the old iron workers.

This drive-to guidebook of historic iron production sites is designed to give the reader a factual

and illuminating look at the people and events that shaped Birmingham into one of America's leading steel centers and, in the process, influenced the national experience.

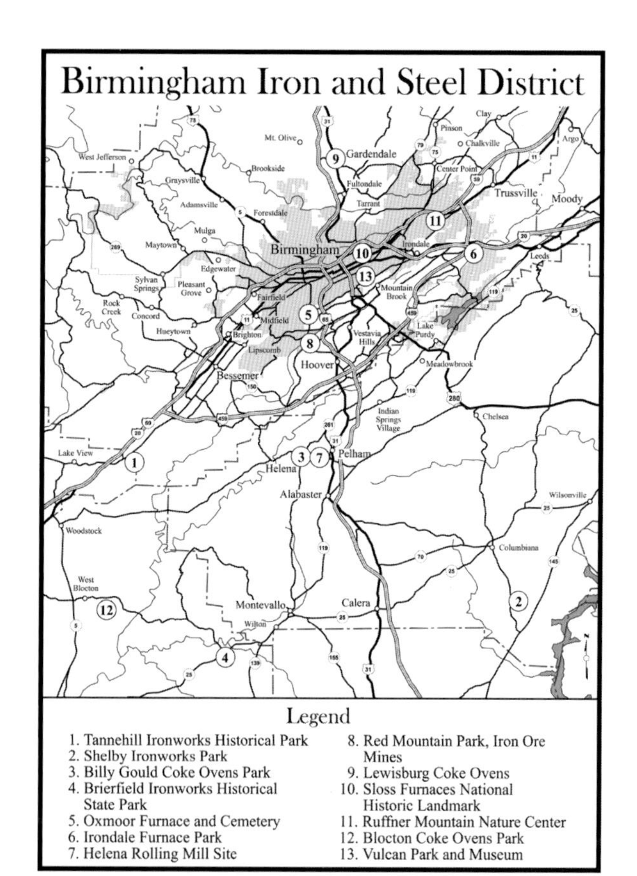
Birmingham Iron and Steel District
Mt. Olive
Gardendale
Pinson
Clay
Chalkville
Argo
West Jefferson
Brookside
Center Point
Fultondale
Graysville
Trussville
Moody
Tarrant
Adamsville
Forestdale
Mulga
Maytown
Birmingham
Irondale
Leeds
Edgewater
Sylvan Springs
Pleasant Grove
Mountain Brook
Fairfield
Rock Creek
Concord
Midfield
Lake Purdy
Hueytown
Brighton
Vestavia Hills
Lipscomb
Hoover
Meadowbrook
Bessemer
Indian Springs Village
Chelsea
Lake View
Pelham
Helena
Alabaster
Wilsonville
Woodstock
Columbiana
West Blocton
Montevallo
Calera
Wilton
Legend
1. Tannehill Ironworks Historical Park
2. Shelby Ironworks Park
3. Billy Gould Coke Ovens Park
4. Brierfield Ironworks Historical State Park
5. Oxmoor Furnace and Cemetery
6. Irondale Furnace Park
7. Helena Rolling Mill Site
8. Red Mountain Park, Iron Ore Mines
9. Lewisburg Coke Ovens
10. Sloss Furnaces National Historic Landmark
11. Ruffner Mountain Nature Center
12. Blocton Coke Ovens Park
13. Vulcan Park and Museum

1

Tannehill Ironworks Historical State Park

Tannehill Ironworks, 1830–1865

> Majestic in the forest, yet ruling no more—it has a burdened, solitary heart.
> —Ethel Armes, 1912

When cotton was king and Jackson was president, one of the earliest iron-making operations in the Birmingham District began an active metal trade along Roupes Creek twelve miles south of Bessemer. It was located near the very spot where Jefferson, Bibb, Tuscaloosa, and Shelby counties then came together. Hillman's Bloomery turned out three hundred pounds of iron per day for agricultural markets in Jonesborough, Elyton, and Tuscaloosa. Its first production was recorded in the fall of 1830. Despite the limited output, Daniel Hillman Sr., an ironmaster from Philadelphia, envisioned a much larger operation. Three decades later, during the Civil War, the complex would grow into one of Alabama's largest producers of iron for Confederate military needs.

Daniel Hillman silhouette, from Hillman Family Bible.

Hillman had been encouraged to come to Alabama by investor Abner McGehee of Montgomery, who acquired rich ore lands in Roupes Valley hoping to find a source of iron for railroads. Seeking his own fortune, Hillman moved from the Hanging Rock Iron Region along the Ohio River, where he had managed the Pine Grove Steam Furnace in Lawrence County, Ohio, and later the Cataract Bloomery on the Little Sandy River in Greenup County, Kentucky. After the death of his wife in 1826, he was anxious to find new iron-making opportunities to build his fortune. With funding from a group of wealthy Alabama planters, including McGehee and Richard B. Walker of Jefferson County, Hillman built his twin-stack bloomery near large beds of brown iron ore just south of Bucksville. He named it the Roupes Valley Iron Company, on a site located in present-

Tannehill Ironworks, 2010 snowfall, Hillman's Forge site is the center structure in the foreground. (Stacey Green.)

day Tannehill Ironworks Historical State Park along Roupes Creek.

The year Hillman built the bloomery, he wrote his son, George: "I believe . . . that my prospects for making a handsome property are better than they ever were during all the course of my life. I wrote to Daniel [another son] and desired him to come to this country for there is one of the best prospects I ever saw for him to make a fortune."

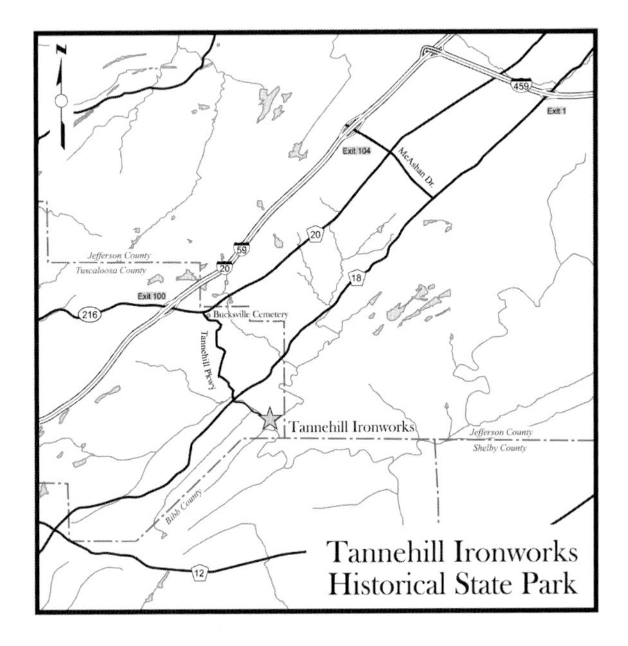

An industrial community grew up around the forge, from which a gently graded wagon road ran to iron ore mines two miles distant. This later became one of the first rail trams in the Birmingham District, although the cars were pulled by a mule. Hillman died in 1832, before he could implement his plans for expansion, including a blast furnace. He is buried in the Bucksville Baptist Cemetery in an unmarked grave.

Noted New York landscape architect Frederick Law Olmsted visited the forge site in 1854 and described a group of ore miners at work: "One [man] was picking a vein having excavated a short audit. The other picked looser ore exterior to the vein. The women and children shoveled out the ore and piled it on kilns of timber where they roasted it to make it crumble." The ore was then carted to the forge where the workers were paid by the load. Olmsted said the mine boss made $3 a day and paid the crew about $1.50 altogether. The mine boss also told Olmsted that the Irishmen who worked at the forge would fight a lot, and after a time, "they get to feel so independent and keerless—like you can't get along with 'em."

After Hillman's untimely death, the forge remained idle until 1840 when it became a part of the nearby Tannehill Plantation and resumed operation. Marion Tannehill, son of the plantation owner, was listed as the "forge master" in the 1850 census. Its owner, Col. Ninian Tannehill, was a veteran of the War of 1812. Beginning in 1836, and with the help of about twenty slaves, he operated a thousand-acre farming and cattle operation including a gristmill, sawmill, and wheat-threshing house.

Today, evidence of Colonel Tannehill's activities can be found in historical records and at ruins of remaining sites. His marriage to Mary Prude on January 28, 1820, was the first on record in the old Elyton Courthouse in Jefferson County, which burned in 1873. The Tannehill family initially lived in Jonesborough and then at Pleasant Hill (McCalla), where he owned a large cotton farm. This farm was near the present-day site of the Colonial Promenade Tannehill shopping mall. Likewise, the site of the Tannehill house is the present location of the Pleasant Hill Methodist Church, just across I-59/20.

The site of Tannehill's second house is located about eight miles to the west near Eastern Valley

Abner McGehee. (Alabama Department of Archives & History, Montgomery, Alabama.)

Col. Ninian Tannehill. (Tannehill Ironworks Historical State Park.)

Moses Stroup. (Tannehill Ironworks Historical State Park.)

John Alexander. (Tannehill Ironworks Historical State Park.)

University of Alabama geology class at Tannehill Furnace site, 1918. (Alabama Department of Archives & History, Montgomery, Alabama.)

Road along the tracks of the Alabama Great Southern Railroad. The house was known for years as one of the grandest residences between Elyton and Tuscaloosa. The railroad, which opened in 1871, was originally called the Northeast & Southwest Alabama Railroad.

In the late 1850s, noted southern ironmaster Moses Stroup, who had made the first iron rails in Georgia, partnered with investor John Alexander to build the first of three blast furnaces at the Hillman forge site. Alexander bought the 280-acre forge track from Tannehill in 1857. Stroup, who sold the Round Mountain Furnace near Centre in 1855, took over management of the Tannehill-Hillman forge and immediately began plans to expand the works.

Furnace No. 1, which could produce six tons a day, went into blast in 1859. Stroup, who had been associated with the Etowah Furnace near Cartersville, Georgia, had moved to Alabama in 1852 to erect the Round Mountain Furnace. While officially known as the Roupes Valley Ironworks, the furnace complex was also referred to as the Tannehill furnaces.

Stroup and Alexander may have been acquainted when both families lived in South Carolina in the 1830s (Stroup's daughter, Sarah, married Alexander's brother, Marshall, and Stroup died at their house near Montevallo in 1878). Court records indicate Stroup, who frequently had money problems, may have defaulted on a lease or sale agreement in 1857 before bringing Alexander in as an investor. When Tannehill sued Stroup in Bibb County Circuit Court, the judge awarded him damages of $2,188.70. To secure payment, the court impounded 3,266 pounds of Stroup's iron.

The Tannehill works influenced the later growth of the iron and steel industry in Birmingham through early experimentation with coke as a fuel source and the first successful reduction of iron ore from Red Mountain in a blast furnace. In 1862, before helping finance new furnaces in Jefferson County, Confederate officials ordered a wagonload of Red Mountain iron ore sent to Stroup's plant at Tannehill to determine whether quality iron could be made from it. Favorable test results paved the way for red ore furnaces to be built in 1863, just south of present-day Birmingham. These were known as the Oxmoor and Irondale furnaces.

Surface cut, Morris Iron Ore Mine at Redding, Red Mountain, Birmingham, from *Harper's Weekly*, September 27, 1890.

With the supply of iron becoming more critical at the start of the Civil War, the Confederacy invested in two additional furnaces at the Tannehill Ironworks, which were erected in 1862. Begun under Stroup's supervision, they were completed by William L. Sanders, a Selma merchant, who assumed control of the iron plant when Stroup was dispatched to help build two new furnaces at Oxmoor in Jefferson County. The Tannehill plant was the only location in Alabama where three blast furnaces operated side by side during the war years.

By 1863 it could produce twenty-two tons of pig iron per day, most of which was shipped to the Selma

Capt. William A. Sutherland, seated center, leader of the Tannehill Furnace raid, 1865. (Ohio Historical Society.)

Arsenal in the form of ninety-pound ingots on the Tennessee & Alabama Rivers Railroad. There it was cast into all the munitions of war. The works at Tannehill were equipped with a steam engine and hot blast stoves. A foundry at the site manufactured eating utensils and hollow-ware for Confederate troops, and a tannery made canteens, harnesses, and leather products for cavalry regiments. During the closing months of the Civil War, the Selma Arsenal and gun foundry, along with a dozen Alabama iron furnaces, were targets of a major federal cavalry raid led by Maj. Gen. James H. Wilson. This raid was the largest cavalry operation of the war.

The Tannehill Ironworks was attacked by three companies of the Eighth Iowa Cavalry on March 31, 1865, under the command of Capt. William A. Sutherland, a veteran of the Battles of Atlanta and Shiloh. Sutherland, who served as adjutant-general to Brig. Gen. John T. Croxton, led his detachment into the furnace yard on the morning of March 31, after heavy rains impeded travel from Elyton. Carefully approaching the place with the fear that Nathan Bedford Forrest's Confederate cavalry might be in the vicinity, the federal raiders moved from the slave quarters to the furnace yard where they encountered a group of female slaves.

George Monlux, a twenty-one-year-old commissary sergeant in Company I, wrote in a report that no employees were in sight when the company arrived at the ironworks, the horses and mules having been taken into the woods. Shots may have been fired at some of those fleeing the premises, accounting for a spent Spencer cartridge found in an archaeological excavation. Only the federal raiders were equipped with seven-shot Spencer repeaters (carbines), one of the most significant upgrades in firepower during the war.

Although the official records indicate three companies were present, Monlux, in his war narrative, lists only two, the third company dispatched perhaps in search of the nearby Williams & Owen Forge. Monlux wrote:

> During the day Company I and Company D were ordered off to the left of the line to burn a large iron works and smelter. I was on the advance of this party and as we rode up to the works there was a large collection of colored ladies in front of a building and one of them addressed me saying, "What are you all quine to do?" I told her we were going to burn the iron works.

Tannehill Ironworks restoration rendering. (Anita Bice, Tannehill Ironworks Historical State Park.)

She replied I am powerful glad of that for it uses up any amount [of my people] every year.

These works were of great size and must have covered more than an acre of ground and near the [furnace] building were tens of thousands of bushels of charcoal. The boys started fires in several places in the building and as it was built of pitch pine the flames spread so rapidly that some of the boys came near being burned before they could get out. We watched the building and these large piles of charcoal until they were well on the way to destruction when we started on to intercept the column.

Croxton's official report was more direct: "Moved at daylight, sending a detachment to the right through Jonesborough to destroy the stores there and three companies of the Eighth Iowa, in charge of Capt. Sutherland, my adjutant-general, to the left six miles to destroy [the] Saunders [*sic*] Iron-Works, which they accomplished, rejoining the column five miles south of Bucksville and ten miles from Trion."

Sanders (Saunders), who took over as furnace master at Tannehill late in 1862 when Stroup went to Oxmoor, apparently expected the attack and had drained his furnaces of their molten metal. Had the furnaces been allowed to chill with hot iron in their chambers, it would have been an industrial calamity as the hardened iron would have shut down the works. While there is evidence shots were fired during the Tannehill Furnace raid, the nearest serious skirmish was at Vance, then called Trion, ten miles to the west.

On April 1, the day after the Tannehill raid, two federal soldiers on horseback attempting to catch up to the moving column en route to attack Tuscaloosa encountered Sam Miller and demanded the gray mule he was riding. Although the owner of the mule had his claim for $175 later rejected by the Southern Claims Commission, he protested loudly that he was a Union man loyal to the government. One of his witnesses was an ex-slave named Mary Owen, who said the mule was taken along "the big road not far from a bridge over the furnace creek near the lands where I was then living." Mary, twenty-six, and her son, Flem, worked for Thomas Hennington

Owen, co-owner of the Williams & Owen Forge, a mile downstream from the Tannehill Ironworks. The "big road" more than likely was the Elyton-Tuscaloosa Road, also known as the old Huntsville Road. This is the route taken by Gen. John T. Croxton's brigade.

The two soldiers may have been a part of the 4th Kentucky, which was held back to bring up the supply wagons. After Sutherland's force rejoined the federal column near Bucksville on March 31, Croxton came in contact with Confederates commanded by Gen. William H. (Red) Jackson about 5 P.M. near Vance. Jackson was en route to join Forrest's main force, which planned to block Wilson's advance toward Selma. With 1,500 men, Croxton had been detached from Gen. Edward M. McCook's division at Elyton on March 30 and sent toward Tuscaloosa to burn The University of Alabama, considered a military school, and the factories there. His other target was the ironworks near the Tannehill Plantation.

As Wilson headed south, burning the Oxmoor and Irondale furnaces, Croxton was ordered to rejoin the column by way of Centreville after he attacked Tuscaloosa. Blocked by Confederates on the move, Croxton never reached the rendezvous point and ended up in Macon, Georgia, on May 1, after both Selma and Montgomery had fallen. In Macon, his detachment became a part of the occupational force. Other units of Wilson's troops captured Jefferson Davis near Erwinville, Georgia, on May 10. Wilson's cavalry operation, which involved more than twelve thousand federal soldiers, set fire to every Alabama

Gen. John T. Croxton. (Library of Congress.)

furnace but one: the works of Hale & Murdock near Vernon in Lamar County off the line of march.

After the war, B. J. Jordan and later James T. Loveless operated the Tannehill Furnace cupola until 1867, using scrap and leftover pig iron for small castings. While a half dozen of the burnt-out Alabama furnaces would be rebuilt, Tannehill was abandoned and reverted to nature. After several proposals to build a new furnace at the site fell through, including

B. J. Jordan. (Miley Collection, Special Collections, Leyburn Library, Washington and Lee University.)

one by the Bethlehem Iron Company of Pennsylvania, the mining of brown iron ore was resumed near Tannehill in 1868 by the Pioneer Mining and Manufacturing Company. The firm later became the Republic Steel Corporation. An article in the *Scientific American* on March 27, 1869, indicated that Pioneer had acquired the old ironworks and planned to "invest $500,000 in them." Those efforts never materialized.

The old furnace ruins were donated to The University of Alabama by Republic Steel in 1952, and

Tannehill double furnace battery built in 1862 with Confederate bonds; photo taken in 1993. (Jet Lowe, Historical American Engineering Record, U.S. Department of the Interior.)

in 1970 they were leased to the Tannehill Furnace & Foundry Commission as the focal point of a new state historical park.

A number of important archaeological investigations have been conducted at the Tannehill site. They included eight different studies beginning with work by The University of Alabama in 1956. The latest such effort, started in 2007, examined the slave quarters area near the furnace complex. Here were found a number of artifacts tied to the plant's slave workforce, which at one time numbered several hundred. The artifacts found included beads, marbles, thimbles, smoking pipes, and eating utensils.

The previously mentioned Spencer carbine cartridge discarded by Sutherland's raiding party was among the items recovered. Other Confederate

Bob Fuller, a Tannehill survivor. (Ethel Armes, 1976.)

military-related items found included a belt buckle, butt plate, and cannonball, indicating the presence of either inspectors from the C.S. Nitre and Mining Bureau or army conscripts assigned to work at the ironworks. These excavations provide a rare glimpse of manufacturing and worker housing at a southern ironworks during the war period.

As part of the American Bicentennial Celebration in 1976, Tannehill Furnace No. 1 (1859) was refired for a run of iron. According to the Smithsonian Institution, it marked the first time in U.S. history that an ironworks out of blast for over a century had been put back into production. The suggestion to

Refire of the Tannehill Furnace, 1976. (Tannehill Ironworks Historical State Park.)

restore the iron furnaces at Tannehill was first made in 1952 by Walter B. Jones, the state geologist, and Fred Maxwell, consulting engineer for The University of Alabama. The Tannehill park board, which had been created by legislative act in 1969, followed

Crowd gathered to watch the Tannehill Furnace refire. (*Birmingham Post-Herald*, E. W. Scripps Company.)

through with the recommendation and invested over $250,000 in restoration work.

Extensive damage had been caused by a dynamite explosion at Furnace No. 1 in 1968. While the culprits were never identified, speculation suggested that hunters or perhaps moonshiners, who did not like the idea of seeing their hideaway become a public park, might have been responsible. The vandalism only intensified efforts on the part of public officials and local citizens to build a memorial to the Alabama iron industry.

The project to put Furnace No. 1 back into blast drew widespread interest from Birmingham District steel mills and foundries. U.S. Steel contributed 50 tons of iron ore, 20 tons of limestone, and 150 tons of coke. American Cast Iron Pipe Company donated

Prof. Ray L. Farabee, furnace master for the Tannehill recharging operation, 1976. (*Tuscaloosa News.*)

an industrial blower and Abex Corporation provided protective equipment. More than 50 volunteers, including workers from U.S. Steel's Fairfield Works, Woodward Iron Company, Koppers Company, Pullman-Standard, Stubbs Foundry, American Cast Iron Pipe Company, and Abex Corp., spent two weeks preparing the furnace for its first run of iron in more than a century.

Tannehill Ironworks today. (Marshall Goggins.)

A crowd estimated by the *Tuscaloosa News* at "more than 15,000" witnessed the event, which produced two and a half tons of pig iron. The metal was cast into replica cannons and souvenir ingots. Dr. Ray Farabee Sr., professor emeritus of chemical and metallurgical engineering at The University of Alabama, served as furnace master.

Professor Farabee, who died in 1994, thought the history of the Tannehill site, along with the successful recharging ceremony, qualified it for national recognition. Partly as a result of his efforts, the site was designated as an American Society for Metals international landmark in 1995. Also recognized by the American Foundrymen's Association, the restored furnaces were placed on the Civil War Trust's National Discovery Trail in 1996 and added to the Alabama Engineering Hall of Fame in 1997.

Somewhat ironically, the Tannehill site may contribute more to the state's economy as a state park than it did in antebellum times as an iron furnace and foundry. Each year more than 450,000 people visit Tannehill Park or attend events there, making it Alabama's largest Civil War–era attraction. Still growing, including an addition to the Iron & Steel Museum in 2008, the Tannehill project has become a national model for how local governments can turn abandoned industrial sites into tourist attractions and outdoor recreational facilities.

Tannehill Today

Tannehill Ironworks Historical State Park

While the ironworks ruins are the focal point of Tannehill Ironworks Historical State Park, surrounding park woodlands spread into Jefferson, Tuscaloosa, and Bibb counties. Park visitors participate in more than thirty major outdoor events, including monthly "Trade Days," which attract three to four hundred vendors from March through November.

In the park center, tourists may view more than forty historic structures, including the largest collection of nineteenth-century log cabins in the South. Tannehill Park is the location of the reconstructed John Wesley Hall Gristmill and the May Plantation Gin House. In 2006, the Alabama Department of Tourism and Travel ranked Tannehill among the top

Iron & Steel Museum of Alabama. (Marshall Goggins.)

three most-visited park and outdoor recreation sites in the state.

The Iron & Steel Museum of Alabama

The Iron & Steel Museum of Alabama, built in 1981, is a regional interpretive center focusing on iron production in the 1800s. Its collections include over ten thousand artifacts, tools, and machines from Alabama iron-making sites, the Alabama Department of Archives & History, the Henry Ford Museum, and the Washington Navy Yard.

The museum displays a number of machine parts uncovered during the eight archaeological

Bloomery Forge exhibit, Iron & Steel Museum of Alabama. (Marshall Goggins.)

investigations. It also features a major collection of artillery projectiles manufactured at the C.S. Naval Gun Works at Selma from 1862 to 1865. In addition, several items recovered from the commerce raider *CSS Alabama*, sunk in 1864 by federal fire off Cherbourg, France, are on exhibit.

A heavy industrial display shed behind the museum features equipment used in Birmingham steel mills from the 1920s through the 1960s. This collection includes a slag pot used at the Ensley Works. When filled with slag it weighed two hundred tons.

A major feature of the museum is a collection of steam engines, including an 1835 Dotterer engine

John Wesley Hall Gristmill, Tannehill Ironworks Historical State Park. (*Tuscaloosa News.*)

once used on a rice plantation near Charleston, South Carolina. It was a part of the Henry Ford collection in the 1920s and was similar to the blast engine used at the Tannehill Ironworks.

John Wesley Hall Gristmill
The reconstruction of the John Wesley Hall Gristmill began along Mill (Cooley) Creek in 1977. The original mill at this site, which operated from 1867 to 1931, burned after milling operations had stopped and Hall, who lived nearby, retired. Dating to Reconstruction, it had replaced an even earlier gristmill

May Plantation Cotton Gin House, Tannehill Ironworks Historical State Park. (Jim Bennett.)

along Roupes Creek near the park entrance built in the 1850s, which was torched by federal raiders during the Civil War. Today, the mill grinds cornmeal for park visitors.

Mill and Roupes creeks, both spring-fed, converge in the park near the iron furnace ruins. A great view of Hall's Mill can be seen from the Tapawingo iron truss bridge (1902) at the entrance to Farley Field. In 1975, the bridge was moved to the park from Turkey Creek, near Pinson.

May Plantation Cotton Gin House

The May Plantation cotton gin house, built in 1858, was moved to Tannehill Park in 1991. Once part of the May family farming operation near Knoxville in

John Scott Young Country Store, Tannehill Ironworks Historical State Park. (Marshall Goggins.)

Greene County, the building includes an 1881 Gullett gin head with feeder and condenser. Cotton tie manufacturers were once a major user of iron produced by Alabama furnaces, including the Shelby Iron Company. The gin house, which utilized mules or horses beneath the building to turn its machinery, is located just behind the Iron & Steel Museum.

John Scott Young Country Store

Relocated from Highway 5 near Brent in 1991, the John Scott Young Country Store opened for business in 1905 and closed in 1941. Today it serves as the park store and campground registration center. It sits across the road from the Camp Ground No. 1

Cane Creek School, Tannehill Ironworks Historical State Park. (Jim Bennett.)

bathhouse. The store sells cornmeal ground at the John Wesley Hall Gristmill, along with grocery and gift items.

Cane Creek School

The Tannehill Learning Center consists of two buildings including the Cane Creek School, a two-room schoolhouse that once served the Cane Creek community near Warrior in Jefferson County.

Built in 1923, it was the successor to one of the first school buildings in Jefferson County on land donated by William Thomas in 1815. The building was moved into the park in 1983.

Kimbrell Methodist Church, Tannehill Ironworks Historical State Park. (Paul Sellers.)

The center also includes the Fowler House School, a converted residence from West Blocton in Bibb County (1860). Learning center programs attract hundreds of Alabama schoolchildren each year.

Kimbrell Methodist Church

The Kimbrell Methodist Church was restored on its mountaintop location in the park in 1972 after being moved from its former site, about a mile away along Eastern Valley Road. Abandoned at the time, it dates to 1905. Today it offers Sunday morning church services for campers and park visitors and is available for weddings. The Green Family Cemetery lies adjacent to the church.

Edwards House, Tannehill Ironworks Historical State Park. (Marshall Goggins.)

The Edwards House

The Edwards House, which serves as the park headquarters, was the home of Trussville's first physician, Dr. John Spearman Edwards. Previously located on the Chalkville Road (U.S. 11), it was moved into the park in 1993 and fully restored. The two-story vernacular-style farmhouse was built in 1879 and had been abandoned for many years.

Oglesby Cemetery

The Oglesby Cemetery, which can be reached along the Iron Haul Trail, contains approximately sixty graves from the 1860s and before. Teamsters drove

Oglesby Cemetery, Tannehill Ironworks Historical State Park. (Jim Bennett.)

wagonloads of Tannehill pig iron past the burial ground on the way to the Alabama & Tennessee Rivers Railroad station at Montevallo during the Civil War. Some of the slaves who worked at the ironworks are believed to be buried here. Once located on the Oglesby farm, which dates to 1858, it is one of the park's most haunting historical attractions.

Park Restaurant

Built in 2005, the current park restaurant replaced an older pioneer-style restaurant building destroyed by a fire in 1999. Built in 1976, the earlier log structure, located along the banks of Roupes Creek, was patterned after the lodge at Judson College in Marion. For many years, it was operated by the Weeks family and was known far and wide for its country fare. The new restaurant building encompasses six thousand square feet of modern dining space and scenic creek

Park Restaurant, Tannehill Ironworks Historical State Park. (Jim Bennett.)

views. The Tram Trail runs beside it, the original route to the ore mines. The restaurant is currently operated by Bob Sykes BBQ.

Tannehill Horse Trails and Stables

The Tannehill horse trails and stables are located near the park entrance on Eastern Valley Road. The operation, under lease to a private company, offers over five miles of riding trails, boarding and riding lessons, and shoeing services. Riding trails retrace old limestone mining sites and run along wooded areas, high ridges, and creeks.

Nineteenth-Century Cabins

More than a dozen original log homes of Alabama pioneers are located in the park, many of which are

Tannehill Stables, Tannehill Ironworks Historical State Park. (Jim Bennett.)

Peel Cabin, Tannehill Ironworks Historical State Park. (Marshall Goggins.)

now used for rental cabins, craft demonstrations, or administrative purposes. They include the Hogan House, which dates to 1834. Relocated from Bibb County, it features a "dog trot" in the center, which cooled the house.

Other nineteenth-century cabins in the park include the Marchant House (Tuscaloosa County, 1871), Dunkin House (Perry County, 1871), Stewart House (Bibb County, 1877), Crocker House (Jefferson County, 1884), Bagley House (Jefferson County, 1856), Thompson House (Bibb County, 1835), and Snead House (Bibb County, 1850). Still others are the Williams House (Bibb County, 1889), Nail House (Jefferson County, 1860), Belcher House (Bibb County, 1870), Collins House (Lamar County, 1870), Wendell Stewart House (Bibb County, 1855), and Peel House (Shelby County, 1890).

The Gott Cabin, a demonstration log house built by noted Appalachian cabin builder Peter Gott of Marshall, North Carolina, was added as part of the "Alabama Reunion" celebration in 1989. Its chimney and porch were taken from the Ash House (St. Clair County, 1850). The Woodward Iron Company Post Office, which dates to 1914, has been converted into a rental cabin, and the Hosmer House (Tuscaloosa County, 1911) serves as a park concession.

Miniature Railway

The park's miniature railway, first opened in 1975, features an 1800s-style engine and riding cars. It travels a mile-long route from near the restaurant

Tannehill train, Tannehill Ironworks Historical State Park. (Marshall Goggins.)

to the gristmill and pioneer farm. In 1997, the new train replaced an older vintage model called the Little Southern.

Park Trails

The park's extensive trails system includes many original roads dating to the Civil War, such as the Grist Mill Trail, the Iron Haul Road, the Slave Quarters Trail, and the Tram Trail, on which ore cars ran to the brown iron ore mines. The trails pass by many historic sites, including the gristmill, the Tannehill furnaces, and the slave cemetery. Several follow scenic creeks.

Federal cavalry raced down the Grist Mill Trail to attack the ironworks on March 31, 1865, as part

Roupes Creek and Cahaba lilies. (Lance Shores.)

Tannehill Park furnace hiking trail near old cabins, Tannehill Ironworks Historical State Park.

of Wilson's Raid. The Civil War–era furnaces like those at Tannehill, Irondale, Brierfield, Oxmoor, and Shelby led to the more modern furnaces in Birmingham, beginning with Alice Furnace in 1880 and the Sloss City Furnaces in 1882.

To reach Tannehill Ironworks Historical State Park follow I-59/20 through Bessemer to exit 100, take a left to Bucksville, and then right on Tannehill Parkway. From I-459 at McCalla turn left on Eastern Valley Road to the park, 10 miles.

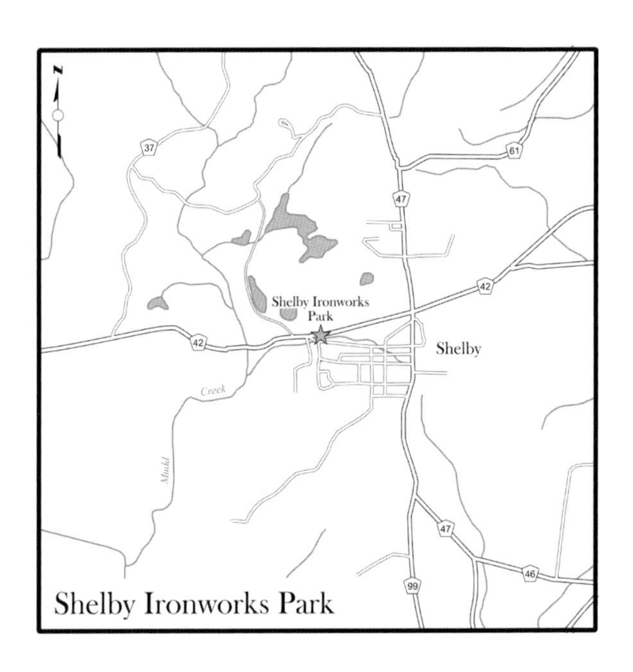
N
37
61
47
42
Shelby Ironworks
Park
Shelby
Creek
Mudd
47
46
99
Shelby Ironworks Park

2

Shelby Ironworks Park

SHELBY FURNACES, 1849–1929

About the time of the great California gold rush, Horace Ware began one of Alabama's great iron enterprises a few miles south of Columbiana in Shelby County. The works would grow into the state's largest ironworks during the Civil War.

Ware's first furnace, twenty-nine feet high, was located a few hundred yards from its brown ore deposits. Unlike many other early ironworks, this plant was operated from the beginning by steam, the first of many innovations here. Not only was it the site of the first large-scale rolling mill in the state in 1860, it was also the first to use the new bell and hopper charging system and hot blast stoves.

A second and larger furnace made of brick was added in 1863 to bolster the Confederate war effort. Armor plate for the *CSS Tennessee* and *CSS Mobile* was rolled here. In the spring of 1864, its operators successfully experimented with coal as a fuel replacement for charcoal, although the practice did not continue.

The furnaces were torched by a detachment of Gen. Emory Upton's Division of Wilson's cavalry on March 31, 1865, but they were rebuilt in 1868

Shelby Furnace, Shelby Iron Company photograph, 1872. (Alex Nuckols Collection, Tannehill Ironworks Historical State Park.)

Shelby Furnace stack. (Copyright *Birmingham News*. All rights reserved.)

and operated through a period of modernization and expansion until 1923. In 1882, John Birkenbine, president of the U.S. Association of Charcoal Iron Workers, called the Shelby Furnace "the queen of American charcoal blast furnaces."

Remaining ruins, including the rolling mill site, are now a part of Shelby Ironworks Park operated by the Historic Shelby Association.

The park is located at 10268 Highway 42, just south of Columbiana, near Shelby Hotel Road. It is twenty-five miles south of Birmingham, a forty-minute drive.

Billy Gould Coke Ovens Park
Buck
Creek
Coke Ovens
Helena
261
52
95
281
91
17
N

3

Billy Gould Coke Ovens Park

Gould and Woodson Coal Mine and Early Coke Ovens, 1860s

One of the earliest identifiable coal mines and coke oven batteries in Alabama can be found along the banks of the Buck Creek at Helena just before it merges with the Cahaba River.

Located in the Riverwoods subdivision, the old coke works may have produced the fuel used in the famed "Eureka Experiment" at Oxmoor Furnace in 1876. This experiment proved coke made from Alabama coal could be used to make good quality pig iron. The success of the venture paved the way for the development of Birmingham into a national steel-making center after the Civil War.

The Helena coke ovens are possibly the oldest yet standing in the United States. They are a part of a planned seventy-eight-acre city park that can be reached off Shelby County Highway 261 at Ruffin Road. They were developed by Billy Gould, who came to Alabama from England as a prospector, miner, and engineer. A partner in the Gould and Woodson mine, he is acknowledged to be the

Billy Gould. (Hugh Allen Gould.)

Helena Coke Ovens. (John Reese.)

first person to make coke from Alabama coal. His partners were Charles and Fred Woodson.

Nineteenth-century coal-mining operations were located in Shelby, Tuscaloosa, and Bibb counties. The Gould Mine was located near the stone piers of the South & North Railroad bridge, a partially built railroad line that extended from Oxmoor Furnace to the Cahaba River and on to Calera during the war. Coal mined in this area beginning in 1862 was estimated at over thirty thousand tons before being put out of operation by Wilson's federal cavalry on April 1, 1865. Ten to twenty years after the war, local residents still appropriated large piles of coal left at Helena and nearby sites.

Though time has destroyed the roof of the coke ovens, their massive stone walls still stretch to their original dimensions, 175 feet in length and 20 feet in width. Its twelve oven openings are intact. The square design of the operation predates the beehive or domed-shaped coke ovens that became commonplace in the 1870s.

To reach the Billy Gould Coke Oven Park from State Highway 261, take Ruffin Road through the Helena City Sports Complex and follow the signs to the park.

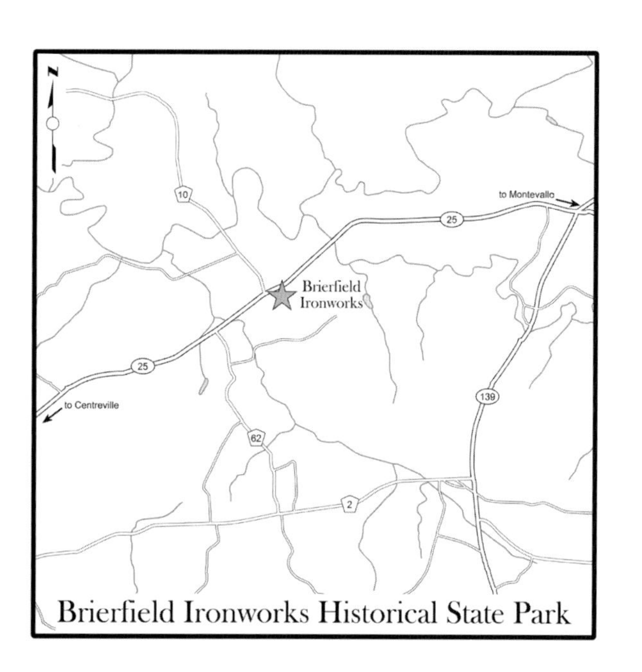
N
10
to Montevallo
25
Brierfield
Ironworks
25
to Centreville
139
62
2
Brierfield Ironworks Historical State Park

4

Brierfield Ironworks Historical State Park

Brierfield Ironworks, 1862–1894

As war clouds loomed on the horizon, the call for expanded Confederate iron production encouraged the construction of new iron furnaces in central Alabama. In the second year of the war, workers at the Bibb Works near Brierfield rushed to complete their first stack.

Under the leadership of C. C. Huckabee and Jonathan Newton Smith, the new plant first served the needs of the local community and later the Confederate army. When the owners refused a government contract to buy all their iron, the government took over the plant and began the construction of an even larger furnace in 1863 under the direction of Maj. W. R. Hunt, chief of the Nitre and Mining Bureau.

The second stack, a large brick furnace forty feet high and equipped with hot blast stoves, became famous for its production of iron for the huge Brooke cannon and iron plates for Confederate gunboats. Most of its iron was sent to the Selma Arsenal

Brierfield Furnace, circa 1900. (Anna Pfaff, Eugene E. Jones Jr. Collection.)

Brierfield Furnace, circa 1910. (Anna Pfaff, Eugene E. Jones Jr. Collection.)

and Naval Gun Works, but war materials were also processed at a large rolling mill at the site.

The Bibb Works, one of the largest in Alabama during the war years, was burned by the 10th Missouri Cavalry under Capt. Frederick Benteen during Wilson's Raid on March 31, 1865. Captain Benteen later commanded one of Custer's units at Little Big Horn in 1876 but survived. The Bibb plant was rebuilt and went back into production under private ownership in 1866. The plant, remodeled several times, produced iron until it closed in 1894. Its ruins can be seen today at Brierfield Ironworks Historical State Park between Montevallo and Centerville off Alabama 25.

Brierfield Park includes several other historic attractions such as the Mulberry Church (1897), the Mahan Cottage (1884), the Ashby Post Office (1908), and the Pratt House (1835). The Mahan Cottage is the remaining portion of the old Brierfield Catholic Church, which was hit by a tornado in 1905. The park also includes a swimming pool, as well as camping, hiking, and meeting facilities.

To reach Brierfield Park from I-65 take Exit 228 west to Alabama 25 approximately 15 miles.

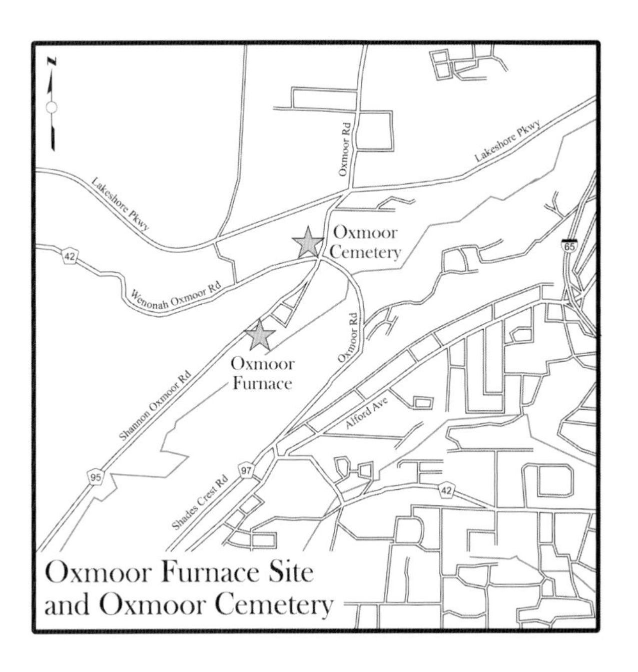
N
Oxmoor Rd
Lakeshore Pkwy
Lakeshore Pkwy
Oxmoor
Cemetery
42
Wenonah Oxmoor Rd
65
Oxmoor
Furnace
Oxmoor Rd
Shannon Oxmoor Rd
Alford Ave
95
97
Shades Crest Rd
42
Oxmoor Furnace Site
and Oxmoor Cemetery

5

Oxmoor Furnace Site

Oxmoor Furnaces, 1863–1923

The rapid growth of Birmingham into Alabama's leading iron-making center had its roots in the Civil War. Following the successful experiment with Red Mountain iron ore at the Tannehill furnaces in 1862, the Confederate government decided to invest in two new blast furnaces in Jefferson County, the first at Oxmoor. Moses Stroup was dispatched from Tannehill to Jefferson County to begin construction of the new plant for the Red Mountain Iron and Coal Company. Red ore, once thought inferior to brown ore, was dug from the Baylis Grace farm near the present-day site of the Spaulding Mine at Grace's Gap. The tracks of the South & North Railroad would run through this natural cut in Red Mountain in 1871.

The works at Oxmoor, originally planned to be two blast furnaces, were built under tight security. William McClane and other furnace builders joined Stroup in a crash project to get the plant in blast by the fall of 1863. Experts from other states were brought in, including Sylvester Bennett from New Orleans as superintendent and Hiram Haines from

Oxmoor Furnace, circa 1872. (Alabama Department of Archives & History, Montgomery.)

Virginia as chief engineer. B. J. Jordan, also from Virginia, was put in charge of the mines.

The entire operation was surrounded by wartime secrecy. Pig iron produced here was shipped to the Selma Arsenal and Naval Gun Works, although a portion also made its way to the Noble Brothers Foundry in Rome, Georgia. As part of Wilson's Raid, Upton's federal cavalry burned Oxmoor Furnace on March 30, 1865, en route to attack Selma.

The plant remained in wrecked condition until it was rebuilt and enlarged in 1873 with financial support from industrialist Daniel Pratt. Both the original stack and the unfinished furnace were enlarged to sixty feet in height by the addition of iron cylinders on top of the stone furnaces.

Oxmoor's coaling yard. Located near the tracks of the South & North Railroad at the Oxmoor Furnace, these charcoal ovens, circa 1872, provided a more efficient fuel reduction process. The Oxmoor Hotel and worker housing are seen in the distance. (Alabama Department of Archives & History, Montgomery.)

Furnace No. 2 was the site of the famed "Eureka Experiment" in 1876 that proved coke produced from Alabama coal could be successfully used to make pig iron. The experiment opened the door to the large-scale growth of the Birmingham Iron and Steel District that followed.

The Tennessee Coal, Iron and Railroad Company acquired the Oxmoor plant in 1892. Although frequently updated, the plant finally shut down in May 1927 and was dismantled. A historical marker along the Shannon-Oxmoor Road marks the site of the first iron furnace in Jefferson County near the present-day Shades Creek Business Park. The old Oxmoor Cemetery is located at the intersection of Oxmoor Road and Ishkooda-Wenonah Road and is now a part of Red Mountain Park.

From I-65/Oxmoor Road intersection in Homewood, travel west on Oxmoor Road to the Oxmoor community approximately 4.5 miles; site is on left at business park.

6

Irondale Furnace Park

Irondale Furnace, 1863–1873

Jefferson County's second furnace had its origins in the demise of the Holly Springs Ironworks in Mississippi, which had produced some of the first cannons in the South for the Confederacy. After the Battle of Shiloh in 1862, it became clear this foundry approximately forty miles southeast of Memphis would fall under federal control.

With help from the Confederate government, its owners purchased a site along Shades Creek in Jefferson County for a new furnace called Irondale (Cahawba). The work, headed by W. S. McElwain, began in the spring of 1863. Extending approximately forty-one feet high, the furnace differed slightly from other furnaces of that era in that it was constructed of heavy masonry at the base and of brick, banded with iron ties, on the mantle.

During the war years, the two Jefferson County furnaces, Irondale and Oxmoor, both used red ore from Red Mountain, a resource largely overlooked by southern iron makers. Sometime during the spring or early summer of 1864, the first practical use of coke for fuel was made at the Irondale Furnace.

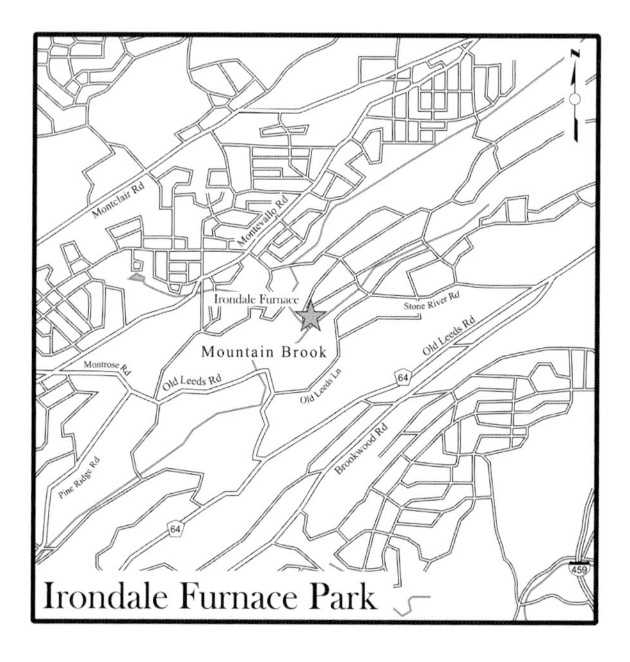

The experiment, conducted at the request of the C. S. Nitre and Mining Bureau in Richmond, proved successful although the test did not persuade the industry to abandon the use of charcoal until after the war.

Along with other Alabama furnaces, the Irondale Furnace was attacked by federal cavalry under Maj. Gen. James H. Wilson (Fourth Iowa Veteran Volunteers) on March 31, 1865, en route to Selma and put out of commission. Attracting northern investment, the plant became the first to reopen after the war, going back into production in 1866. McElwain's rebuilt furnace was enlarged and converted to steam blast. After a series of ownership changes, it was shut

Irondale Furnace, circa 1873. (George Gordon Crawford Collection, Special Collection, Samford University, Birmingham, Alabama.)

Irondale Furnace remains, 2007. (Jim Bennett.)

down during the Panic of 1873. Today, its ruins are the focal point of a city historical park in Mountain Brook at the corner of Stone River Road and Old Leeds Lane.

The old Irondale Furnace commissary, circa 1840s, is located on Montevallo Road at Glenbrook Drive (see historical marker); furnace ore mines are located behind present-day Trinity Medical Center on Montclair Road and several blocks to the west in an area known as the Helen Bess Mines.

From Old Leeds Road in Mountain Brook, travel left on Old Leeds Lane to Stone River Road.

7

Helena Rolling Mill Site

Helena Rolling Mill, 1864–1923

The Helena Rolling Mill, also known as the Central Iron Works, was classified as "top secret" when built along Buck Creek in 1864. Constructed to manufacture war materials for the Confederacy during the Civil War, it was razed during Wilson's Raid on March 30, 1865.

One of six rolling mills in the state in operation during the war years, it had been the work of Montgomery merchants Hannon, Offutt & Company. The mill's first superintendent was Thomas S. Alvis, a Virginia ironmaster who had previously built another rolling mill for the Confederate government at Selma.

The Central Rolling Mill, which had a daily capacity of ten tons, had just begun operation when attacked by federal cavalry in Upton's division. After lying abandoned for seven years, Rufus W. Cobb, B. B. Lewis, and Richard Fell bought the property at a tax sale and organized the Central Iron Works Company. They repaired the war-damaged building and machinery and began operations again in 1873. The mill's capacity was expanded to fifteen tons per day.

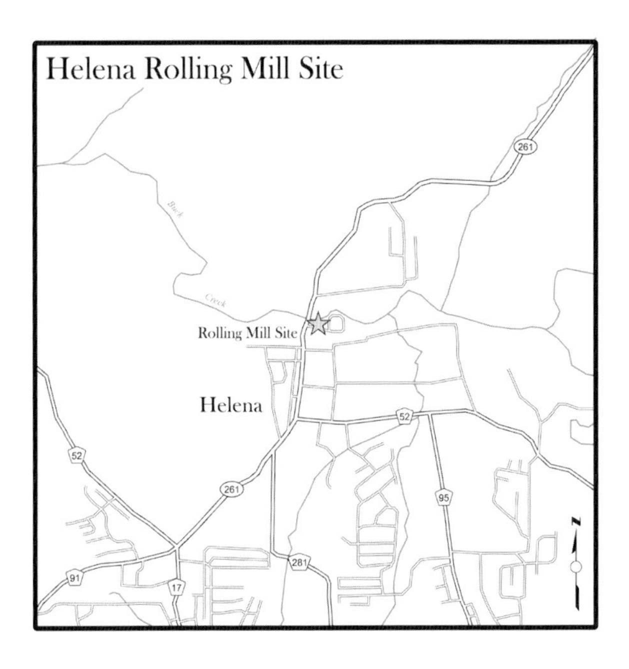

Helena Rolling Mill, also known as the Central Iron Works, circa 1875. (Ken Penhale.)

Shelby Rolling Mill, circa 1890. After the Central Iron Works went under, the new Shelby Rolling Mill Company expanded to eight thousand tons per year in merchant bar, band iron, and light T-rails. The old dam is still in place. (Shelby County Historical Society Museum and Archives; Bobby Joe Seales.)

By 1875 the works were almost exclusively devoted to making iron ties or bands for baling cotton, a trade in which they cornered the local market. In 1883 the plant began manufacturing nails in an effort to make a profit in declining economic times. In 1888, however, the Central Iron Works went bankrupt. A new company called the Shelby Rolling Mill acquired the property in 1889 and remodeled the plant to manufacture merchant bar, band iron, and light T-rails. Its chief officers were George H. Dudley, E. A. Hopkins, and Richard Fell Jr. Within three years, this company would also fail.

A succession of companies would occupy the site including the Alabama Tube & Tire Company in 1901, which made wrought-iron pipe, and the

Conners-Weyman Steel Company in 1908, which produced hoops, light bands, and, once again, cotton ties. The plant was dismantled in 1923 after nearly sixty years of iron-casting operations. A historical marker designating the rolling mill site is located at the intersection of Lake Davidson Lane and Route 261 in Helena.

From Route 261 toward Helena from Hoover, turn left on Lake Davidson Lane; see historical marker at milepost 1.2.

8

Red Mountain Park, Iron Ore Mines

Red Mountain Mines, 1880s–1960s

As one of the nation's largest urban parks, the Red Mountain Park and Greenway stretches four and a half miles along the crest of Red Mountain, the site of major iron ore mines in the early 1900s. Larger than Central Park in New York City, Red Mountain Park covers 1,200 acres of mining property formerly owned by U.S. Steel Corporation and the Woodward Iron Company. Its conversion into a public park represents the largest philanthropic gift in U.S. Steel's history.

The Red Mountain mines first opened in 1863 with the construction of the Oxmoor Furnace, a major source of Confederate iron during the Civil War. The old rail routes are now used as hiking trials. The mines were legendary as the source of the Birmingham District's red iron ore deposits. Mineral resources from the mountain fed local furnaces for almost one hundred years until the mines closed in 1962.

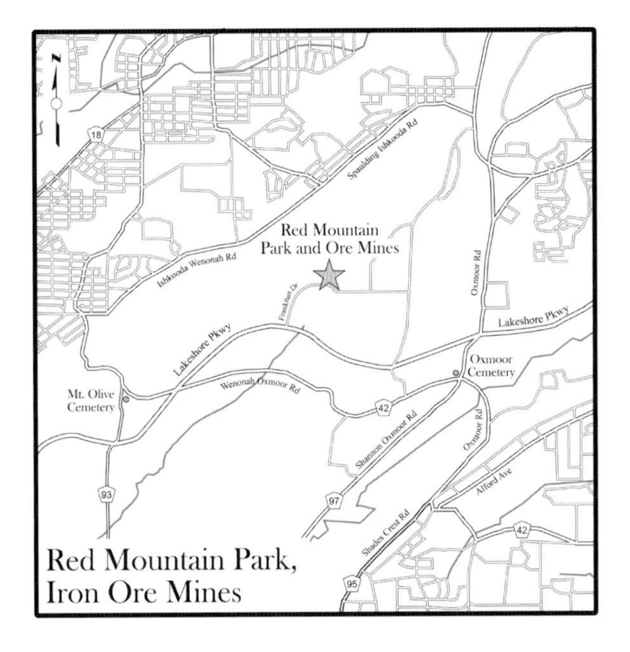

Red Mountain Park,
Iron Ore Mines

The park property includes twenty-one historical sites including four concrete slope portal entries, the Redding vertical shaft, and several rough-slope mines and the associated rail line routes that led to U.S. Steel, Republic, and Woodward Iron Company plants in Jones Valley. A dramatic view of the major iron ore seam on Red Mountain can be seen from the Red Mountain Expressway cut between Birmingham and Homewood.

A master plan for development, including an underground mining museum, is being implemented with access initially limited to hiking, birding, picnicking, and sightseeing activities. The welcome and exhibit center is being designed to focus

Begun in the 1880s, this red ore mine atop Red Mountain supplied the Alice Furnaces, the first iron furnace built in Birmingham. It was operated by the Woodward Iron Company from 1912 to 1927, which had two shafts at the site, one slope, another vertical. The mine is among a dozen major mining sites being preserved in Red Mountain Park. (Eric McFerrin.)

Hoist house, Redding Mine, 1917, Woodward Iron Company, Red Mountain Park. (Eric McFerrin.)

on the region's iron-mining heritage and the growth of Birmingham as a major industrial center.

Like Tannehill and Brierfield, Red Mountain Park is a public property owned by the State of Alabama. It is operated by the Red Mountain Greenway and Recreational Area Commission. The park is located just north of Lakeshore Parkway and west of I-65, four miles southwest of downtown Birmingham. It is expected to open on a limited basis in 2010–11.

From Lakeshore Drive, turn right onto Sydney Road (south entrance); from Ishkooda-Wenonah Road, turn left at Wilson Road (north entrance). The entrances may be under construction. Public hikes and bike rides are scheduled using temporary entrances.

9

Lewisburg Coke Ovens Park

LEWISBURG COKE OVENS, 1880s–1950s

In the 1880s, Jefferson County's Five Mile Creek area exploded with coal mines and mining camps. Uniquely situated, the Birmingham Industrial District was one of the only places in the world where the necessary components for iron and steel production—iron ore, limestone, and coal—were centrally located. Built circa 1889, three rows of beehive-type ovens wrapped around the Lewisburg hillside in North Jefferson County.

In 1901 the coal mines and coke ovens of the Lewisburg district owned by the Jefferson Coal, Iron & Railroad Company were acquired by the Alabama Consolidated Coal & Iron Company. Coke produced here was used at the Etowah Furnaces in Gadsden, Ironaton Furnaces near Talladega, and the Williamson and Sloss furnaces in Birmingham.

In 1916, the major coal mine in the area was known as the Mary Lee Mine.

Beehive coke ovens, made of brick, are heated to temperatures above 2000°F for thirteen to eighteen hours, at which time the hot gases are driven off. To produce coke, coal is crushed, washed, and baked to

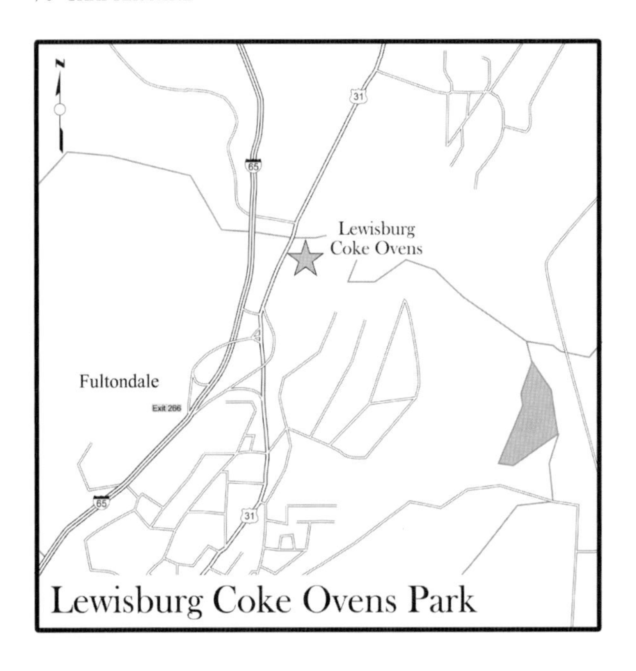

Lewisburg Coke Ovens, Lewisburg Coke Oven Park. (Cawaco RC&D Council, Inc.)

Mary Lee Mine entrance. (Birmingham Public Library Archives, Cat. #1556.16.68.)

remove impurities, leaving a nearly pure form of carbon that is useful not only as a fuel in iron and steel production but also for foundries. Workers fed batteries of coke ovens from top-loading railcars filled with coal, which was very labor-intensive work.

In the mid-1920s as steam power gave way to electrification, the Sloss-Sheffield Steel and Iron Company, then the owner of the Lewisburg ovens, abandoned much of the old equipment, preferring to use their modern North Birmingham by-products facility. Though dormant for a few years, the coke ovens were pressed back into service during World War II only to fall once again into inactivity after the war. Coke production in the area ended in the 1950s.

The coke oven battery remained hidden in a blanket of vegetation until Darryl Aldrich, building inspector for the City of Fultondale, led an effort to save them in connection with the twenty-eight-mile-long Five Mile Creek Greenway project. Today the oven remnants are incorporated into Lewisburg Coke Oven Battery, a Fultondale city park.

Take I-65 North to Exit 266. The old mine site is just off Ellard Road.

10

Sloss Furnaces National Historic Landmark

Sloss Furnaces, 1882–1970

> This magic little city of ours has no peer in the rapidity of its growth. . . . its permanent mountains groaning to be delivered of their wealth. . . . the El Dorado of iron-masters.
>
> —Colonel James Powell

Following the Civil War, Alabama's economy began to shift away from the dominance of agriculture. In 1871, prominent Alabamians joined forces to form the city of Birmingham with the intention of exploiting the mineral resources of north-central Alabama, where every ingredient necessary for making iron could be found within a thirty-mile radius. One of these men was James Withers Sloss, a north Alabama merchant and railroad man.

Sloss was born in Mooresville, Alabama, in 1820. His father, Joseph, emigrated from County Derry, Ireland, to Lexington, Virginia, with his parents in 1803. Joseph later moved to Tennessee, where he met

Postcard, Sloss Furnaces. (Sloss Furnaces National Historic Landmark.)

Steam locomotive with Sloss in background. (Sloss Furnaces National Historic Landmark.)

and married Clarissa Wasson from Alabama. They relocated to Mooresville to farm and raise a family.

James, the Sloss family's oldest son, became an apprentice bookkeeper for a local butcher at age fifteen. At the end of his seven-year term, he married a local girl, Mary Bigger, and used his savings to buy a small store in Athens, Alabama. By the 1850s, James Withers Sloss had extended his mercantile interests throughout northern Alabama and eventually became one of the wealthiest merchant and plantation owners in the state.

In the early 1860s, realizing the need for the expansion of southern rail lines, Sloss became active in railroad construction. As Birmingham journalist Ethel Armes noted in a 1910 newspaper article, "Sloss had a voice in county and state politics, and was taking up the fight for railroads with vigor, and that good Irish tongue of his to boot." In 1867, following years of negotiations, all rail lines between Nashville and the Tennessee River were consolidated into one company, the Nashville and Decatur, with James W. Sloss as its first president.

Sloss not only promoted the expansion of southern rail lines but also became a chief proponent of Alabama's postwar industrial development, notably the area around present-day Birmingham. As a local newspaper stated years later, "His influence will be found connected with every important industrial and commercial enterprise in the State during the latter half of the nineteenth century." Sloss became allied with the strongest railroad in the South: the Louisville and Nashville, known as the L&N. He realized that the L&N had reached a critical stage and needed to find an outlet to the Gulf of Mexico to access new markets and connections with Montgomery and Mobile. Sloss forged an alliance between the L&N and the Elyton Land Company (the company eventually responsible for developing the mineral district of Jones Valley and the city of Birmingham).

In 1868 the Elyton Land Company had started construction on the South & North Railroad and had reached the Jones Valley area when they shut down because of financial problems. James Sloss

Elyton Land Company, 1883. (Birmingham-Jefferson History Museum Collection.)

traveled to Louisville and presented the president of the L&N, Albert Fink, with a glowing picture of the mineral riches of the Jones Valley area and the future rail traffic it was capable of generating. He offered to lease the Nashville and Decatur to the L&N if that line would assume the Elyton Land Company's railroad debts, pay interest on its bonds, and complete work on the gap between Decatur and Birmingham. When completed, the L&N could run the combined system all the way from Nashville to Montgomery. Fink liked the idea, and he and his partners invested more than $30 million in furnaces, mines, wharves, steamship lines, and other Alabama operations, eventually reaching Mobile. By 1888 the L&N was hauling iron, coal, and other mineral products, outweighing the nation's entire cotton crop.

The L&N transformed Birmingham from a squalid jumble of tents, shanties, and boxcars into a thriving

Terminal Station with Magic City sign. (Birmingham Public Library Archives, Cat. #OVH 85.)

community. From the beginning, Sloss had no doubt that the city was destined for greatness. Because of its close proximity to raw materials, cheap labor, ideal climate, and newly developed rail connection to the rest of the country, Birmingham industrialists were confident that iron could be made more cheaply in Jones Valley than anywhere else. While speaking at the Elyton Land Company's annual meeting in 1873, James Powell, an early Birmingham promoter, coined the community's nickname (the Magic City) by referring to "this magic little city of ours."

In 1878, determined to tap the rich mineral areas surrounding Birmingham, Henry DeBardeleben, son-in-law and heir to the Pratt cotton gin fortune, joined forces with Sloss and Truman Aldrich, a successful mining engineer, to form the Pratt Coal and Coke Company, the first large coal company in Alabama. Pratt soon became the largest mining enterprise in the district. In 1882, the Pratt Company sold out to Memphis entrepreneur Enoch Ensley, who invested nearly $1.5 million and increased mining operations to a daily output of 2,500 tons. By 1885, Ensley's Pratt Company owned 70,000 acres of coal lands, 710 coke ovens, and 30 miles of railway. With over a thousand free and convict laborers, the Pratt mines produced "first-class coking coal" for pig iron furnaces throughout the Birmingham area. Tennessee Coal and Iron (TCI) acquired the company in 1886 and immediately added additional mines to the Pratt holdings. Since 1907, the mines have been the property of U.S. Steel.

Another southern entrepreneur crucial to the development of Birmingham's industrial district was Henry DeBardeleben, son of an Autauga County planter and heir to the Pratt fortune. Daniel Pratt had migrated to Georgia from New Hampshire in 1821, and in 1883 built the world's largest cotton gin in Prattville, Alabama. Orphaned at ten, Henry DeBardeleben became Daniel Pratt's legal ward. After serving in the Confederate army and fighting at Shiloh, he returned home to run a bobbin factory owned by Pratt and married his daughter, Ellen.

In 1879, inspired by the success of the Pratt Company, DeBardeleben created Alice Furnace, named for his oldest daughter. The largest blast furnace erected in Alabama to that date, it was also one of the first (along with Edwards Furnace at Woodstock), designed to use coke. Sixty-three feet high and fifteen feet wide at the bosh (hottest part of the furnace), Alice Furnace achieved an average daily production

Birmingham in the 1900s with Sloss in background. (Birmingham Public Library Archives, Cat. #1556.13.90.)

of fifty-three tons of pig iron under the supervision of Thomas T. Hillman, a veteran ironmaster whose grandfather had built the Roupes Valley Forge in 1830. Known as "Little Alice," it set off a wave of furnace construction that was later referred to as the "Great Birmingham Iron Boom."

During the 1880s, as pig iron production in Alabama rose from 68,995 to 706,629 gross tons, nineteen blast furnaces were built in Jefferson County alone. DeBardeleben's 1880 "Little Alice" was the first, and James Sloss built the second, Sloss City Furnaces, in 1882. DeBardeleben agreed to supply Sloss with coking coal, and Mark W. Potter, who owned red ore deposits on Red Mountain, agreed to supply the ore. Sloss took the contracts to Louisville and won the financial backing of E. D. Standiford, president of the L&N. Following his return to Birmingham, Sloss and his sons, Maclin and Frederick, filed papers at the Jefferson County Courthouse in the spring of 1881 to incorporate the Sloss Furnace Company.

Construction of Sloss's new furnace began in June 1881, when ground was broken on a fifty-acre site that had been donated by the Elyton Land Company. Harry Hargreaves, a European-born engineer, was in charge of construction. Hargreaves had been a pupil of Thomas Whitwell, a British inventor who designed the stoves that would supply the hot-air blast for the new furnace. Sixty feet high and eighteen feet in diameter, Sloss's new Whitwell stoves were the first of their type built in Birmingham and were comparable to equipment used in the North. Birmingham industrialists were impressed that much of the machinery used by Sloss's new furnace would be manufactured in the South.

This southern-made machinery included two blowing engines and ten boilers, and was constructed in Birmingham at Linn Iron Works, a company started in 1885 by Finnish-born entrepreneur Charles Linn. A local reporter described them as the "largest engines ever made south of Pittsburgh." Although the enormous boilers were too complex to be built in Birmingham, they, too, were made in the South. Walton & Company, located in Louisville, completed the task in time for the opening in 1882. In its first year of operation, Sloss Furnace Company sold 24,000 tons of iron. At the 1883 Louisville Exposition, the company won a bronze medal for "best pig iron."

Sloss photo, 1880s. (Sloss Furnaces National Historic Landmark.)

James W. Sloss not only exported his iron but managed to supply large amounts to local agriculture for items such as traps, pipes, and stoves. The majority of Sloss pig iron, however, ended up in major cities such as Cincinnati, Louisville, St. Louis, Nashville, Chicago, Detroit, and Cleveland. Sloss's iron production had an economic edge, as pig iron costs in northern plants averaged $18.30 per ton in 1884 while pig iron in the South could be produced for $10–$11 a ton. The foundry product that stimulated by far the most growth of Sloss-Sheffield and the emergence of the southern foundry trade in general was cast-iron pipe, used for transporting water, sewage, chemicals, fuels, gases, and other substances. Two general classes of pipe were produced: soil pipes

Early furnace workers. (Sloss Furnaces National Historic Landmark.)

(through which water or other substances flowed by force of gravity) and pressure pipes (through which substances were pumped by force).

As Birmingham flourished, so did Sloss. He retired in 1886 and sold the company to a group of financiers who guided it through a period of rapid expansion. The company reorganized in 1899 as Sloss-Sheffield Steel and Iron, although it would never make steel. With the acquisition of furnaces and extensive mineral lands in northern Alabama, Sloss-Sheffield became the second largest pig iron company in the Birmingham District. Company assets included seven blast furnaces, 1,500 beehive coke ovens, 120,000 acres of coal and ore land, five Jefferson County coal mines and two red ore mines, quarries in North Birmingham, brown ore mines in Russelville, and 1,200 tenements for its workers.

Ladle advertisement from *Pig Iron Rough Notes* magazine. (Sloss Furnaces National Historic Landmark.)

In 1914, at the start of World War I, Sloss-Sheffield was among the largest producers of pig iron in the world. By 1917, when America entered the war, military service claimed approximately 700 Sloss-Sheffield employees. In the same year, Sloss-Sheffield's total assets of $27 million ranked 28th among primary metal manufactures in the United States and 206th among all industrial firms.

In October 1925 the directors of Sloss-Sheffield, realizing the need to match northern technological standards (and increase production), authorized $625,000 toward a massive rebuilding program. Early in 1926, the #2 blast furnace was completely dismantled and reconstructed; the following year, the #1 blast furnace was rebuilt. Each of the new furnaces had an average daily capacity of 400–450 tons,

Production shot of Sloss. (Sloss Furnaces National Historic Landmark.)

compared to a previous maximum output of about 250 tons. These changes helped carry Sloss-Sheffield through America's worst economic crisis, the Great Depression, which began in 1929.

In 1933, the United Mine Workers launched a drive to organize Alabama's coal, iron, and steel industries. Union organizers, labeled by the local press as communists and outside agitators, were threatened, beaten, and run out of town. Nonetheless, Sloss-Sheffield ore miners were the first in the state to affiliate with the union, forming Local 109 on July 17, 1933. Because of unionization, Sloss-Sheffield was forced to raise wages and grant fringe benefits. In 1936, the company began giving a one-week paid vacation to all employees who had worked for the company for five years or more.

In 1938, President Franklin D. Roosevelt described the American South as "the nation's number one economic problem." The Great Depression had been especially damaging to industry, but in 1939, World War II resurrected the South's economy. Under wartime regulations, Sloss-Sheffield fell under the jurisdiction of the War Production Board's iron and steel branch, which was charged with coordinating the output of approximately three thousand foundries scattered throughout the United States. As Sloss-Sheffield switched from peacetime to wartime operation, its cast iron was used in fragmentation and incendiary bombs, as well as mortar shells, marine engines, and hand grenades. About 60 percent of all hand grenades used by American forces in the war came from the Birmingham area. Other strategic and defense-related items that were made by cast iron foundries ranged from air compressors, brake drums, and camshafts to soil pipes for army bases. Although airplanes and ships were made largely of aluminum and steel, they could not exist without cast-iron machine parts used in aircraft manufacturing. Sloss cast iron also helped supply food for wartime needs through its use in agricultural equipment, cookware, and containers.

From 1941 to 1945, almost all of Sloss-Sheffield's pig iron went into munitions. A sign in front of City Furnaces stated, "This is a U.S. Arsenal." And by the end of the war, more than five hundred Sloss employees had seen military duty.

When America entered the war, Birmingham's population surpassed 226,000; Jefferson County's

African American workers. (Sloss Furnaces National Historic Landmark.)

was more than 431,000. Nearly half the labor force was employed by the iron and steel and mining districts; more than two-thirds of the industries' workers were African American, and the industrial workplace was rigidly segregated until the 1960s. At Sloss, for example, men bathed in separate bathhouses, punched separate time clocks, and attended separate company picnics.

The company operated as a hierarchy. At the top there was an all-white group of managers, accountants, engineers, and chemists; at the bottom were all-black labor gangs, partly recruited until 1928 by the use of forced convict labor in the mines. In the middle, a racially mixed group performed a variety of skilled and semi-skilled jobs. Even in the middle group, however, white workers held the higher-

Sloss Bessie Mines. (Sloss Furnaces National Historic Landmark.)

paying, higher-status positions as stove tenders, boiler makers, carpenters, and machinists. Black workers were restricted to such "helper" roles as carpenter helper, machinist helper, and stove tender helper.

As Birmingham's population exploded in the late nineteenth century, Sloss Furnaces, Tennessee Coal and Iron Company, and numerous other furnaces and mines built low-cost housing throughout the Birmingham area. Sloss Quarters, the forty-eight houses adjacent to the site, were designed for African American workers pouring in from the depleted lands of the Black Belt. They were typical shotgun-style structures, with two rooms set on foundation posts and no indoor plumbing in the early years. Until the 1930s, drinking and cooking water came from a faucet placed at the end of each row of houses.

Rain barrels caught water for the laundry. Nonetheless, many of these homes proved an improvement over sharecropper shacks.

Housing in the quarters served two purposes: it attracted family men, thus lowering the rate of absenteeism, and made available a supply of labor for emergencies. It also made workers dependent on the company. Sloss realized that hiring workers with wives and families and furnishing houses at reasonable rents helped promote stability. Sloss charged $4 per month for a one-room dwelling, $5 for two rooms, and $6 for three.

The industrialists of the Birmingham District found it difficult to keep workers on the job; turnover and absenteeism rates were high. Providing housing for workers in company towns was a means of keeping the workforce available for labor and retaining the services of skilled men.

Although the new housing was an improvement over their dilapidated rural dwellings, life for Birmingham's black iron and steel workers and their families was far from idyllic. Despite the drawbacks, the quarters provided a relatively cohesive community setting. Watermelon cuttings, barbecues, chitlin suppers, quilting bees, baseball games, and church suppers were a few of the social events organized by the women of Sloss Quarters. And with Thomas School nearby, black children had access to educational opportunities almost unheard of in the rest of the South.

Sloss Quarters was not a company town in the strictest sense because it did not provide a company

U.S. Pipe, Sloss. (Sloss Furnaces National Historic Landmark.)

school or recreational services. There was, however, a doctor's office, as well as a commissary that became the focal point of life in the quarters. The company store, or commissary, served as both pay office and shopping center. Shoes, cloth, tools, seed, appliances, and a variety of foodstuffs could be purchased. In the early days workers paid for items with company script called "clacker." After the use of clacker declined, families continued to purchase goods at the commissary, but more often they bought on credit. Although some Sloss workers found the credit system helpful, others saw it as a way for the company to take advantage of its workers.

Sloss Quarters was dismantled in the late 1950s as maintenance and repairs became a drain on the company's resources. At the same time, higher wages, improved transportation, and environmental con-

Sloss in disrepair. (Sloss Furnaces National Historic Landmark.)

cerns encouraged residents of the quarters to seek better housing away from the plant.

On September 12, 1952, Sloss-Sheffield merged with U.S. Pipe & Foundry and Claude Lawson, president of Sloss-Sheffield, became the new president of U.S. Pipe. While the new headquarters was being built, a major modernization program, the first Sloss-Sheffield had undertaken since 1931, got under way in North Birmingham. The blast furnace at this site was the largest and most modern installation ever built in the United States for foundry pig iron. Built by 500 workers who spent an estimated 1 million man-hours in its completion, it towered 225 feet above ground level. The hearth, 25 feet in diameter, had a capacity of 1,000 tons per day. It was lined with 638,000 fire bricks, weighing about 2,400 tons.

Although it climaxed nearly eight decades of development since the founding of the Sloss Furnace Company in 1881, the blast furnace's construction was ill timed. Even as it was being built, the American iron and steel industry began to face severe foreign competition. Some came from West German plants that had been destroyed by Allied bombing attacks during World War II and rebuilt under the Marshall Plan. In addition to being ultramodern, these plants had lower labor costs than older American installations. Meanwhile, foreign plants were also exporting low-cost iron and steel to American markets that were no longer protected by high tariffs. The market was also changing. Aluminum was beginning to replace iron in automobile engine blocks. Iron stoves, once a mainstay of the industry, were no longer made. Steam radiators had gone out of use.

The main cause of the disaster that overtook American merchant pig iron producers, however, was what has been referred to as a "technological discontinuity." Suddenly, gray iron was obsolete. In 1948, two Americans and an Englishman developed a new type of cast iron by adding cerium of magnesium to the mix. Unlike gray iron, the new alloy, known as ductile iron, contained graphite in the form of tiny spheroid nodules instead of flakes. While retaining some of gray iron's desirable features, including a low melting point and easy machinability, ductile iron had greater compressive strength. Combining various features of cast iron and steel, it became the material of choice.

Sloss worker at tap. (Sloss Furnaces National Historic Landmark.)

In addition, foundries now used increased quantities of scrap steel in their mixes because its carbon content was significantly lower than that of pig iron. This reduced demand for foundry pig iron at the same time that foreign pig iron became cheaper than the America product. The result was disastrous for Sloss Furnaces.

While Sloss fought to survive the industrial changes that were taking place throughout the country, Birmingham experienced its most climactic decade. By 1960, nine decades of racial injustice had produced problems in Birmingham's industrial district. Three out of every four whites were skilled, but only one of six blacks could claim such status. "Operation New Birmingham" began communicating across racial lines to institute a new social order that gradually

Sloss hot blast stoves. (Sloss Furnaces National Historic Landmark.)

improved the city's tarnished image. And by the mid-1960s, Clarence Dean, an African American worker at Sloss, had finally secured the promotion to iron pourer that he had wanted so long. But his career as a furnace worker was nearly over. By 1970, the last furnace of its kind in the Birmingham area, Sloss Furnaces, would go out of blast forever.

After Sloss stopped production, Birmingham's character changed dramatically. Throughout the decade the air in Birmingham, once polluted and dangerous, improved enormously, and the natural beauty of the city's surroundings became apparent. By the mid-1970s, Birmingham won *Look* magazine's All-American City Award. Birmingham was now better known as an educational and medical center than as an industrial city. By the 1980s, Birmingham's population was composed of professionals, not blue-collar workers. This was due not only to educational and medical institutions but also to the city's large number of communications, electronics, engineering, and insurance firms. Birmingham's involvement in the growing trend toward professional and service-based occupations was typical of a postindustrial economy.

While Birmingham and the world around it changed, questions remained about what to do with the idle Sloss Furnaces. In July 1972, the site was placed on the National Register of Historic Places, but no funds were available for development. The homeless slept on the grounds at night, vandals removed brass and copper parts, and the plant became a public disgrace. Civic leaders talked about converting it into a theme park, but nothing was done because the eighteen-acre location was too small and the projected $65 million cost too high.

In 1976, the Historic American Engineering Record (HARE) conducted a detailed survey of the property, jointly funded by the city council and the National Parks Service, to assess its historical significance and prepare a permanent architectural record. In the spring of 1977, the city of Birmingham earmarked $3.3 million to refurbish the furnaces and turn them into an industrial museum.

View from #1 Sloss Furnace. (Sloss Furnaces National Historic Landmark.)

In 1977, Jim Waters, an architect who became president of the Sloss Furnace Association (an organization of individuals from all backgrounds interested in preserving the site), outlined a plan to restore the site, increase public accessibility, and renovate the employee bathhouse for use as offices and a visitors center.

For more than sixteen months, sandblasting and painting, grading and seeding, paving and lighting,

Stokin' the Fire BBQ and Music Festival. (Sloss Furnaces National Historic Landmark.)

removal or repair of hazardous structures, renovation of the bathhouse, and creation of displays proceeded. In May 1981, the tract became one of the eighty-seven sites in the United States designated as a National Historic Landmark.

For Birmingham, Sloss is important and significant because of what it represents. There is not a single major historical issue in the city's first one hundred years that is not directly related to Sloss: the importance of railroads, an economy built on heavy industry, company towns, and segregation. The very character of the city is traceable to heavy industries like Sloss.

Sloss helps preserve a Birmingham that has all but disappeared.

Today, Sloss not only is dedicated to preservation and education but also serves as a center for community and civic events, including rock concerts, theater performances, music festivals, barbeque events,

Sloss metal arts pour. (Sloss Furnaces National Historic Landmark.)

weddings, and reunions. Public presentations, lectures, and site tours provide insight into Sloss's industrial heritage.

Sloss houses one of the best innovative and active metal arts programs in the nation. The program offers workshops and classes that address all aspects of creating metal sculpture: pattern making, mold making, casting, welding, cutting, and forging. This program is rooted in Birmingham's connection to iron and steel and replicates one hundred years of industrial processes in Birmingham.

Education programs allow visiting school groups to design and create their own cast iron tiles using trivet-sized sand molds. Students watch while Sloss artists melt scrap iron in a small furnace and fill the completed sand molds. As the metal cools, it hardens into a cast-iron tile and within moments each student takes away an original work of art. Outreach programs focusing on the lives of workers visit local schools and impart a sense of life in a 1900s industrial site, what it was like to work in a blast furnace, and how these facilities affected the growth of Birmingham.

Since its opening in 1982, Sloss has been actively collecting oral histories. These histories are from black and white workers who speak not only about the technological aspects of the plant but what life was like working, as one worker put it, at the "gates of hell." Oral histories from women residing in Sloss Quarters provide insight into the day-to-day activities in company housing. While their husbands labored in the surrounding mines and blast furnaces (and literally built the city of Birmingham), and while faced with segregation and discrimination, these remarkable women built strong family foundations by adapting their customs and traditions to their new surroundings.

Sloss Furnaces is more than a landmark or museum; it represents the character and spirit of the South's industrial heritage. As one former worker stated, "A lot more than iron flowed from those furnaces. Our whole culture did. Our whole way of life."

From I-59 take the 22nd Street exit. Turn left onto 1st Avenue North. Cross the viaduct (Sloss Furnaces is on the right). Turn left onto 34th Street. Turn left onto 2nd Avenue North. Turn left onto 32nd Street North. Turn right into the Sloss Furnaces entrance under the viaduct.

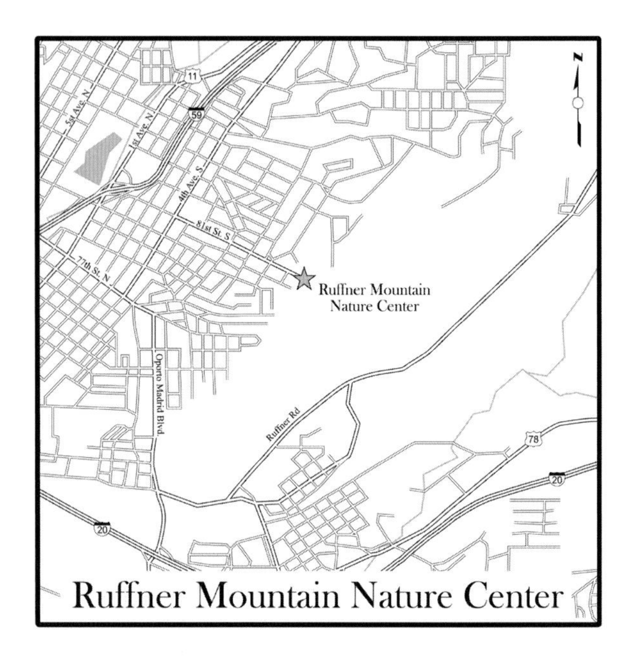

5st Ave. N
1st Ave. N
4th Ave. S
81st St. S
77th St. N
Oporto Madrid Blvd.
Ruffner Rd
Ruffner Mountain
Nature Center
11
59
78
20
20
N
Ruffner Mountain Nature Center

11

Ruffner Mountain Nature Center

In 1883, William Ruffner, acclaimed geologist at Washington and Lee University in Virginia, published a detailed geological survey of the rich mineral resources along the route of the Georgia Pacific Railroad from Atlanta to the Mississippi River through the Birmingham District. In 1887, John W. Johnston, president of the Sloss Furnace Company, one of Birmingham's largest producers of pig iron, opened limestone quarries and soft ore mines in the Irondale area, one of the sites mentioned along Ruffner's detailed route. Johnston's decision to purchase the rich mineral lands proved wise. Daily output of the Irondale mines in the first year of production was 135 tons. In 1893, the Sloss Furnace Company expanded operations by working outcrop and harder ores and by the turn of the century, daily output for all three of Sloss's Irondale mines (now named Ruffner mines) was 200 tons.

By 1900, Birmingham had become the leading center of southern iron production. Before long, the area became a magnet for thousands of workers seeking jobs and a better life for their families. As

Ruffner Quarry, Ruffner Mountain Nature Center. (Bob Farley/f8Photo.)

Birmingham's population exploded, Sloss Furnaces, Tennessee Coal and Iron, and numerous other furnaces and mines built low-cost housing throughout the industrial district. Company housing was a part of the Sloss mining operation at Ruffner from its initiation in the late 1800s. Housing was under the supervision of the mine superintendents who usually lived at the site. The houses were sold to employees in the 1950s and most have since disappeared.

Although in 1929 Sloss discontinued its Ruffner operations, in 1952 the company installed a plant at Ruffner to beneficiate (crush and separate) ore before transporting it to its furnaces in downtown and North Birmingham. The mines permanently closed in June 1953.

Today, Ruffner Mountain is a 1,011-acre nature preserve operated by the Ruffner Mountain Nature Coalition. The preserve includes eleven miles of marked trails (once the old mining roads) and a

Ruffner Building, Ruffner Mountain Nature Center. (Bob Farley/f8Photo.)

visitor center containing native Alabama plants and animals. Founded initially with only twenty acres in 1977, Ruffner is currently the second-largest urban nature preserve in the United States. Expansion plans are currently under way and include an environmentally innovative "Green" Tree Top Visitor's Center, an educational pavilion, and the restoration of the Ruffner Wetlands.

From Interstate I-59, take exit #132 at First Avenue North. At the end of the exit ramp, bear to the right onto 1st Avenue. Go to the first traffic light at 83rd Street and take a right. You will go back under the interstate and at the first stop sign at 2nd Avenue South take a right. Go two blocks and take a left on 81st Street South.

12

Blocton Coke Ovens Park

Blocton Coke Ovens, 1888–1904

Built in the heart of the Cahaba Coal Field in 1888 to process coal from mines operated by the Cahaba Coal Mining Company, the Blocton "beehive" coke ovens were among the largest in Alabama at the time.

Here Truman H. Aldrich, a mining engineer who had organized the Cahaba Coal Mining Company in 1883, constructed four batteries totaling 467 coke ovens capable of producing 600 tons daily. While first used to provide fuel for the Woodstock Iron Furnaces in Anniston owned by stockholders Samuel Nobel and A. L. Tyler, coke produced at Blocton later found its way into Birmingham and Bessemer foundries and steel mills including the Oxmoor and Trussville furnaces.

Coke is one of three ingredients needed to make pig iron, the source of steel; the other ingredients are limestone and iron ore. Coal was top-loaded into the Blocton coke ovens and heated to 2,800 degrees to burn off the impurities, leaving coke, which is almost pure carbon. After it was heated, it was dragged out, quenched with water, then loaded into railcars.

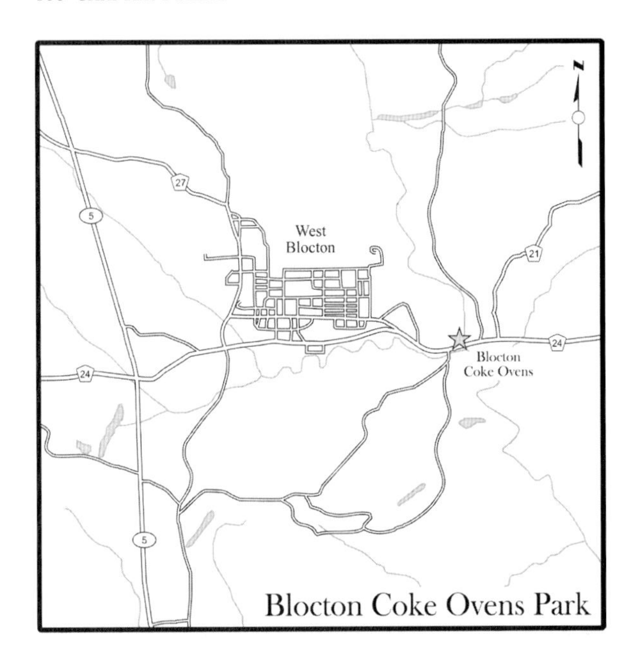

Blocton Coke Ovens, circa 1900, Tennessee Coal, Iron and Railroad Company. (TCI Annual Report, 1910.)

Blocton Coke Ovens Park. (Rusty Crouthers.)

Ten coal mines operated in the area, the last opened by the Tennessee Coal, Iron and Railroad Company (TCI) in 1915. TCI acquired the property in 1892 and used coke produced here at its plants in Birmingham. The Blocton Coke Ovens operated for about twenty years before being phased out in 1904 by newer technology that recovered by-products.

As the mining community expanded to over 3,600 residents, the towns of Blocton (which no longer exists) and West Blocton grew up nearby. Today the remains of the coke ovens form the central attraction at Blocton Beehive Coke Ovens Park, located about a quarter of a mile from West Blocton traveling east on County Highway 24, also called Cahaba River Drive.

With help from the United Mine Workers of America, Blocton Coke Ovens Park opened on a limited basis in 1996. The town of West Blocton as well as volunteers with the Cahaba River Society plan to transform the land where the ovens are located into a west Alabama tourist attraction that will extend to the Cahaba Wildlife Refuge along the Cahaba River.

From West Blocton, go east on Highway 24 approximately ¼ mile.

13

Vulcan Park and Museum

At the beginning of the twentieth century, the industrial boomtown of Birmingham had become the largest city in Alabama and one of the fastest growing in the nation. Local mines were producing over 4 million tons of coal and 1.5 million tons of iron ore. Iron foundries were prevalent, and by 1900 Alabama was producing more than 25 percent of the nation's foundry iron—overtaking Pennsylvania as the number-one producer in the United States.

Barely thirty years old in 1903, Birmingham was thriving and ready for recognition. The opportunity to express its civic pride materialized in 1903 when Alabama was invited to sponsor an exhibit at the 1904 St. Louis World's Fair. When the state government declined to support the project, the Commercial Club of Birmingham (which later became the Chamber of Commerce) stepped in to organize and finance the exhibit. Two of the club's leading members, Frederick M. Jackson and James MacKnight, suggested a colossal iron statue symbolizing Birmingham's industrial might and Alabama's supremacy in iron production.

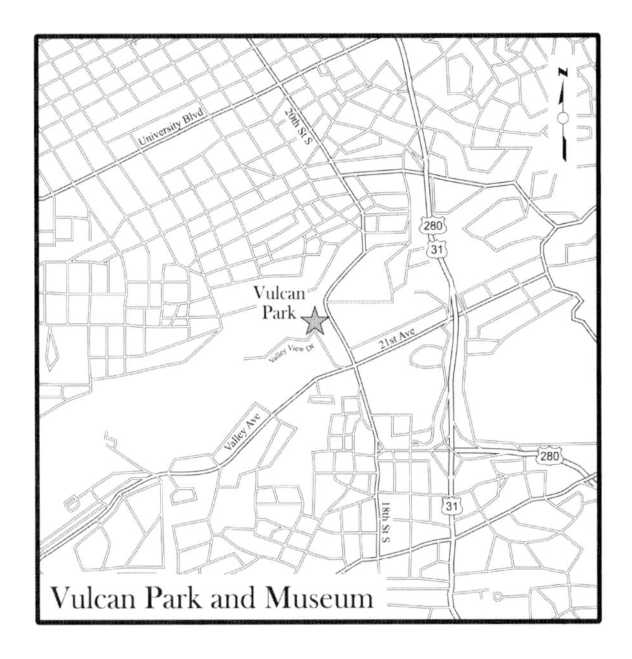

Vulcan Park and Museum

The choice was not surprising since large statues were frequently commissioned for homes and public parks, and especially for world fairs. Frederic Bartholdi's Statue of Liberty, for example, was completed in 1886. For their symbolic statue, the Commercial Club chose Vulcan, the Roman god of the forge, a working man usually pictured in sooty surroundings and engaged in hard labor. Giuseppe Moretti, an expatriated Italian sculptor residing in New York, was commissioned to design the statue. The casting was to be done by the Birmingham Steel and Iron Company's new foundry at 1421 First Avenue.

Only Vulcan's feet and lower legs were in place when the St. Louis World's Fair opened on April 30,

Vulcan at the state fair, Vulcan Park and Museum. (O. V. Hunt, Birmingham Public Library Archives, Cat. #1556.21.68.)

1904, but by the dedication in early June, the 55-foot tall, 120,000-pound Vulcan stole the show. Viewed by twenty million visitors at the fair's enormous Palace of Mines and Metallurgy, Vulcan was named the official Alabama exhibit, and an international panel of jurors awarded the statue the grand prize. MacKnight, Jackson, and Moretti received silver medals for their part in the project. Following the closing of the fair in 1904, Vulcan was disassembled and brought back to Birmingham, free of charge, by the L&N Railroad.

In December 1935, TCI deeded 4.45 acres of land to the city of Birmingham for Vulcan Park. Assisted by the Kiwanis organization, numerous Italian immigrant stonemasons, and the state's Works Progress

Vulcan, Vulcan Park and Museum. (Rob Lagerstrom.)

Administration, Vulcan was finally reassembled on Red Mountain in 1936.

Today, standing high above the valley where he was created in 1904, Vulcan, the largest cast iron statue in the world and largest metal statue ever made in the United States (55 feet, 120,000 pounds), stands as a symbol of Birmingham's rich industrial heritage.

After a four-year renovation, the new 10-acre Vulcan Park and Museum reopened to the public in 2004 and welcomed over 100,000 visitors its first year. The Vulcan Center Museum provides the centerpiece of the park's educational mission and tells the story of Birmingham's past, present, and promise for the future.

Traveling in the direction from Huntsville, follow I-65 South to exit 261A onto I-59 North/I-20 East toward Gadsden/Atlanta. After a short distance, take exit 126A onto Highway 280 East. Travel 2.8 miles and exit at 21st Avenue South. Turn right onto 21st Avenue South and travel .4 miles. Vulcan Park and Museum entrance is on your right immediately upon crossing the four-way intersection of 21st Avenue and Richard Arrington Boulevard.

Glossary

alloys Metallic substances composed of two or more elements possessing properties different from those of their components.

battery A grouping of furnaces, forges, or coke ovens.

beehive coke oven A structure resembling a beehive designed to reduce coal to coke for use as fuel in a blast furnace; lined with refractory brick and charged from the top, it also utilizes a spray of water. The coke is removed by tearing out a temporary wall covering an arched front or from a door built for the purpose. Beehive ovens were about 12 ½ feet in diameter and seven feet high. They were capable of producing about 3.1 tons of coke from five tons of coal.

blast The air forced into the bottom of an iron furnace to augment the combustion of the fuel source, charcoal, or coke.

blast furnace A chambered iron-making apparatus using iron ore, limestone, and charcoal or coke in which air, hot or cold, is forced into the bottom to enhance reduction.

bloomery An early iron plant that usually used charcoal for fuel and a forced air process powered either by a steam engine, water wheel, or water trompe. The last iron produced in America in a bloomery was in 1901. In early terminology, a bloomery was essentially a Catalan forge raised in height to produce more iron.

blowing engine Machinery producing air blast to expedite ore reduction in an iron furnace, usually a steam engine. Blowing tubs could also be operated from a water wheel.

boiler A cast iron tank, probably riveted, circa 1800s, in which water was converted into steam that operated a blowing engine; now made of low carbon steel.

bosh The widest portion of a blast furnace, frequently the middle; sometimes used to denote the entire core of the furnace.

brown iron ore Limonite, basic iron ore of early Alabama furnaces, the iron content of which varies from 52 to 66 percent iron.

carbine A type of short-barreled rifle, including the 54-caliber 7-shot Spencer repeating carbine used by U.S. Cavalry during Wilson's Raid into Alabama in 1865.

cast iron A first-generation iron usually used for kettles, pipes, or heavy castings; containing a large amount of carbon; it is brittle and not suitable for fine moldings or tools.

charcoal A dark, porous form of carbon made by partially burning wood in an airless kiln for use as a furnace or foundry fuel, circa 1800s. All Alabama furnaces used charcoal as the primary fuel source through the Civil War.

coal A black, combustible, carbon sedimentary rock used as furnace fuel, usually converted to the distilled form of coke; Alabama coal is identified as bituminous.

coke A solid, coherent cellular residue of the destructive distillation of coal processed by high temperatures, usually in coke or beehive coke ovens; used as a fuel source in blast furnaces replacing charcoal. Well established, especially in northern states during the Civil War, coke furnaces by 1869 outproduced charcoal furnaces by 553,341 tons to 392,150. In the next decade coke became the primary fuel source for American iron furnaces.

cupola A type of blast furnace in which the metal and the fuel are in immediate contact, similar to early American furnaces; also refers to the remelting of pig iron and scrap to make iron suitable for foundry use. A cupola furnace to remelt iron for casting purposes was first used in France around 1700.

Eureka Experiment An experiment conducted at the Oxmoor Furnace near Birmingham in 1876 that proved good quality pig iron could be made from coke produced from Alabama coal.

gray iron Unrefined cast iron so named for its characteristic gray fracture; uses may range from heavy machine parts to skillets.

hoist house The elevator mechanism that pulls by cable underground mining cars carrying ore, coal, or workers.

hollow-ware Iron cast into vessels including skillets, ovens, bowls, pots, and cups that have a significant depth and volume.

hoops Circular metal bands for holding materials together such as staves of a barrel.

ingots Elongated blocks of iron usually weighing about 90 pounds made by pouring molten iron into sand molds of a standard size to ship to finishing mills or ironworking shops. Ingots at Tannehill ranged from 18 to 33 inches long and were about 3 ½ inches deep; also called pigs. Ingots from U.S. Steel's Fairfield Works in the 1960s weighed 12 tons.

iron bands Metal binders that encircle brick construction to make it stronger or to hold other material together.

iron cylinder Circular, barrel-like, metal extensions that increased the capacity of iron furnaces made of stone or brick in the 1800s.

iron ore Earthen material in its natural state containing iron oxides; Alabama ore is mainly of the brown and red varieties, extracted in either surface or deep-slope mines.

iron ties A metal binder used to bind bales of cotton weighing five hundred pounds; may also refer to the clip connecting the two ends of metal bands.

light T-rails An early form of railroad iron used for track.

merchant bar Metal shapes such as angles, channels, flats, rounds, squares, and strips (but not reinforcing bars) that foundries shape into different products.

molten metal Metal in liquid form from a furnace or cupola used in casting operations for making products or ingots.

munitions War supplies, ammunition, shot, shell, and projectiles.

pig iron Iron that has been run directly into "pigs" or sand molds from a blast furnace; cast iron. Pig iron casting machines were in use in the late 1890s.

red iron ore Hematite, iron ore type that replaced brown ore as the basic resource in the Alabama iron industry, the iron content of which is the equivalent of 70 percent.

refiring The ignition of a fuel source to put an iron furnace back into operation after a shutdown.

rolling mill A mill that rolls iron into various shapes including slabs or finished products such as rails, beams, or plates. Common in England in the early 1700s, rolling mills used rolls that were smooth-surfaced iron cylinders that pressed hot metal into thin, flat sheets. In 1783 Henry Cort took out a patent for a mill with grooved rolls, making it possible to roll iron into finished shapes.

soft ore mines In soft-ore mines extraction was done by shovel and pick, later by mechanical scrapers. Where the ore was harder in deep-slope mines, powder blasts were necessary.

stack A stone, chambered structure resembling a truncated pyramid used in early to mid-nineteenth-century iron furnaces where ore reduction took place by applying heat and air blast.

steam blast Usually refers to a forced air flow from a steam engine that is channeled into the bottom of an iron furnace to enhance ore reduction. The air may be preheated in stoves.

steam engine A heat engine that performs mechanical work using boiling water to produce mechanical motion. Invented three hundred years ago, these devices made the Industrial Revolution possible. The first commercially successful engine did not appear until 1712; it was invented by Thomas Newcomen and updated earlier designs.

steel Iron subjected to intense heat and mixed with up to 2 percent carbon to give it added hardness and strength; also sometimes mixed with alloys.

trivet A hot plate placed between a serving dish or bowl and a dining table, usually to protect the table from heat damage; trivet also refers to tripods used to elevate pots from the coals of an open fire.

wrought iron A malleable iron put together from pasty particles low in carbon and not requiring fusion. Production reached its peak in the 1860s when used in ironclad warships and railways.

References

Alabama Ironworks Source Book. McCalla, AL: Alabama Historic Ironworks Commission, 2006. http://www.alaironworks.com/.

Armes, Ethel. *The Story of Coal and Iron in Alabama.* 1910. Reprint, Birmingham, AL: Book-Keepers Press, 1972.

Bennett, James R. *Historic Birmingham and Jefferson County.* San Antonio: Historical Publications Network, 2008.

———. *Tannehill and the Growth of the Alabama Iron Industry.* McCalla, AL: Alabama Historic Ironworks Commission, 1999.

Camp, J. M., and C. B. Francis. *The Making, Shaping, and Treating of Steel.* Pittsburgh: United States Steel Company, 1951.

Confederate Citizens Group, Red Mountain Iron and Coal Company, NARA M346. Letters from Sylvester Bennett to Col. James Burton, Montgomery, AL, September 29, 1862, March 19, 1863, National Archives, Washington, DC, 1862–63.

Fisher, Douglas Alan. *The Epic of Steel.* New York: Harper and Row, 1963.

Iron Age: Handbook of Terms Commonly Used in the Steel and Nonferrous Industries. Philadelphia, 1971.

King, Edward. "The Great South." *Scribner's Monthly* vol. 8, no. 5 (September 1874): 523.

Lewis, David W. *Sloss Furnaces and the Rise of the Birmingham District.* Tuscaloosa: University of Alabama Press, 1994.

Moldenke, Richard. *Library of Iron and Steel: Principles of Iron Founding.* Vol. 3. New York: McGraw-Hill, 1917.

Morris, Philip A. *Vulcan & His Times.* Birmingham: Birmingham Historical Society, 1995.

Newcomb, Ellsworth, and Hugh Kenny. *Miracle Metals.* New York: G. P. Putnam's Sons, 1962.

Pool, J. Lawrence, and Angeline J. Pool. *America's Valley Forges and Valley Furnaces.* Dalton, MA: Studley Press, 1982.

Ruffner Mountain Nature Center. http://www.ruffnermountain.org/.

Southern Claims Commission. Claim of Samuel Hamaker, Jefferson County, AL, December 5, 1875, pp. 28–31. Rejected: U.S. House of Representatives, National Archives, Washington, DC, NARA M1407. Claim filed April 12, 1872.

Walker, James H. Jr. *Roupes Valley.* Bessemer, AL: Montezuma Press, 1991.

White, Marjorie. *The Birmingham District: An Industrial History and Guide.* Birmingham: Birmingham Publishing Company, 1981.

Woodward, Joseph H. II. *Alabama Blast Furnaces.* Birmingham: Woodward Iron Company, 1940.

Index